Chromium: Metabolism and Toxicity

Editor

Desmond Burrows, M.D., F.R.C.P.
Department of Dermatology
Royal Victoria Hospital
Belfast, Northern Ireland

CRC Press, Inc.
Boca Raton, Florida

Library of Congress Cataloging in Publication Data

Burrows, Desmond.
  Chromium—metabolism and toxicity.

  Bibliography: p.
  Includes index.
  1. Chromium—Metabolism. 2. Chromium—
Physiological effect. 3. Chromium—
Toxicology. I. Title. [DNLM: 1. Chromium—
Metabolism. 2. Chromium—Toxicity.
QU 130 B972c]
QP535.C7B87   615.9′25532   82-4512
ISBN 0-8493-5447-1   AACR2

This book represents information obtained from authentic and highly regarded sources. Reprinted material is quoted with permission, and sources are indicated. A wide variety of references are listed. Every reasonable effort has been made to give reliable data and information, but the author and the publisher cannot assume responsibility for the validity of all materials or for the consequences of their use.

All rights reserved. This book, or any parts thereof, may not be reproduced in any form without written consent from the publisher.

Direct all inquiries to CRC Press, Inc., 2000 Corporate Blvd., N.W., Boca Raton, Florida, 33431.

© 1983 by CRC Press, Inc.

International Standard Book Number 0-8493-5447-1

Library of Congress Card Number 82-4512
Printed in the United States

PREFACE

Chromium is important in medicine because of its widespread use (about 400,000 tons are processed in the U.S. each year), toxicity, carcinogenic effect, and its high sensitization index in low concentrations. In the U.S., the use of chromium is divided as follows: (1) Metallurgical industries—57%; 50% chromite oxide, mainly trivalent, primarily used alloyed with iron, nickel, and cobalt; 60% being stainless steel. (2) Refractory materials 30%; 34% chromic oxide mainly trivalent used in furnace bricks because chromite has a high melting point (2040°C)—this use is declining as open hearths are replaced by basic oxygen furnaces. (3) Chemical Industry 13%, 45% chromic oxide, tanning agents, pigments, catalysts, plating, and wood preservers. Chromium therefore has a very widespread use in industry and the following workers could be in contact with chromium:

Abrasive makers
Acetylene purifiers
Adhesive workers
Airplane sprayers
Aircraft workers
Alizarin makers
Alloy makers
Aluminium anodizers
Anodizers
Antifreeze makers
Army (uniform)
Artificial flower makers
Automobile workers

Battery makers
Biologists
Blueprint makers
Boiler scalers
Brass cleaners
Brewers
Bricklayers
Builders laborers

Candle makers
Carburetor cleaners
Cement workers
Central heating workers
Ceramic workers
Chemical workers
Chromate workers
Chromium alloy workers
Chromium allum workers
Chromium platers
Color television makers
Copper etchers
Copper plate strippers
Corrosion inhibitor workers
Crayon makers

Diesel locomotive repairmen
Donkey greasers
Drug makers
Dye makers
Dyers

Electroplaters
Enamel workers
Engine cleaners
Explosive makers

Fat fishfryers
Fat purifiers
Fireworks makers
Flypaper makers
Food laboratory workers
Foundry workers
Furniture polishers
Fur processors

Glass fiber makers
Glass frosters
Glass makers
Glass polishers
Glue makers
Gramophone makers

Histology technicians
Housewives—detergents

Ink makers
Jewellers
Joiners

Laboratory workers
Laundry workers
Leather finishers
Leather glove makers

Linoleum workers	Porcelain decorators
Lithographers	Pottery frosters
Machinists	Pottery glazers
Magnesium treaters	Printers—making plates and off set printing
Magnetic tape makers	
Margarine workers	Railroad engineers
Match makers	Refractory boiler cleaners
Metal cleaners	Refractory brick makers
Metal etchers	Rubber makers
Metal polishers	
Metal workers	Sheet metal workers
Milk preservers	Shingle makers
Milk testers	Silk screen makers
Miners	Soap makers
Mordanters	Sponge bleachers
	Stainless steel and other chrome alloy workers
Oil drillers	
Oil purifiers	
	Tanners
Painters	Textile workers—dyes, mordant, and cutting markers
Paint makers	
Palm oil bleachers	Tin can labelers
Paper dyers	Tire fitters
Paper manufacturers	
Paper waterproofers	Ventilation engineers
Pencil makers	
Perfume makers	Wallpaper printers
Photoengravers	Wax workers
Photographers	Welders
Plasterers	Wood preservative workers
Platinum polishers	Wood stainers

In a book such as this it is almost inevitable that there will be some slight overlap as disciplines impinge on each other. This has been kept to a minimum except where it was felt useful to have the same topic dealt with from a slightly different viewpoint. A good deal of thought has been given to whether it is more correct to use the term "chromium" or "chromate," particularly in the Skin and Allergy section. The point is often made that the element chromium itself, seldom, if ever, causes toxic effects or sensitization. However, in Chapter 5, the case is made for the use of "chromium" in sensitization and so in this section chromium is used, whereas in the skin section "chromate" is used because in effect it is only chromate salts of chromium which are harmful to the skin.

I would like to thank the following who gave their helpful advice on the content and detail of the book: Dr. J. Martin Beare, Dr. Charles Calnan, Professor Harold Elwood, Professor Sigfrid Fregert, Dr. James L. Gardiner, and to Colin Burrows for manuscript correction. I am very grateful to Mr. W. D. Linton and Miss M. Saunders, Queen's University of Belfast Medical Library for their help with bibliographic citation.

THE EDITOR

Desmond Burrows (Consultant Dermatologist, Royal Victoria Hospital, Belfast) first became interested in chromate dermatitis when working as a MRC Research Fellow with Professor Calnan at the Institute of Dermatology, London from 1960 to 1961. Since that time, he has published numerous papers on clinical aspects of chromate allergy as well as contributing to symposia on the subject. He has contributed chapters on contact dermatitis to *Dermatology*, Dr. P. Hall-Smith, Ed., Butterworth and to *Industrial Dermatitis*, Dr. A. Griffiths, Ed., Blackwell. He is a Fellow of the Royal College of Physicians, Edinburgh and Ireland; Member of the Executive Committee, British Association of Dermatologists and British Contact Dermatitis Group; and Chairman of the Dermatology Committee, Central Committee for Hospital Medical Services, British Medical Association. He has been Visiting Professor to Oregon Health Sciences Center, Portland and Stanford University, Palo Alto, Calif.

CONTRIBUTORS

P. Lesley Bidstrup, M.D., F.R.C.P., F.R.A.C.P.
Sloane Terrace
London, England

Desmond Burrows, M.D., F.R.C.P.
Consultant Dermatologist
Royal Victoria Hospital
Belfast, Northern Ireland

Sverre Langård, M.D., M.Sc.
Department of Occupational Medicine
Telemark Sentralsjukehus
Porsgrunn, Norway

Andrew H. G. Love
Department of Medicine
Institute of Clinical Science
Royal Victoria Hospital
Professor of Gastroenterology
Dean of the Faculty of Medicine
Queen's University of Belfast
Belfast, Northern Ireland

Ladislav Polak, M.D., C.Sc.
Scientific Specialist
F. Hoffmann-La Roche & Co., Ltd.
University Lecturer
Dermatological Clinic
University of Basel
Basel, Switzerland

TABLE OF CONTENTS

Chapter 1
Chromium—Biological and Analytical Considerations 1
G. Love

Chapter 2
The Carcinogenicity of Chromium Compounds in Man and Animals 13
S. Langard

Chapter 3
Effects of Chromium Compounds on the Respiratory System 31
P. L. Bidstrup

Chapter 4
Immunology of Chromium 51
L. Polak

Chapter 5
Adverse Chromate Reactions on the Skin 137
D. Burrows

Index 165

Chapter 1

CHROMIUM—BIOLOGICAL AND ANALYTICAL CONSIDERATIONS

A. H. G. Love

TABLE OF CONTENTS

I. Introduction 2

II. Biological Aspects 2
    A. Occurrence 2
    B. Absorption 3
    C. Tissue Levels 3
    D. Serum Values 4
    E. Excretion 4
    F. Environmental Exposure 4

III. Analytical Techniques 5
    A. Methods 5
        1. Optical Spectroscopy 5
        2. Activation Analysis 6
        3. Mass Spectroscopy 6
        4. X-Ray Emission Techniques 7
    B. Choice of Analytical Method 8
        1. Accuracy and Precision 8
        2. Limit of Detection 8
        3. Elemental Coverage 8
        4. Determination of Chemical Form 9
        5. Spatial Distribution of Elemental Composition 9
        6. Practical Considerations 9
    C. Future Developments 9

References 10

I. INTRODUCTION

Of the 92 elements, from hydrogen to uranium, to which the student of biology and medicine was introduced in his early training, only a few remain of general or constant significance in his later work. The discoveries and refinements of organic and biochemistry have perhaps (until recent years) diverted attention from the importance of inorganic chemistry in biological sciences. One area in which this has been evident is the biological significance of metals and their salts. For centuries, some metals have been recognized as industrial poisons, largely due to the high and easily recognized concentrations occurring in the environment. At such times, biological occurrence of these metals was referred to in such dismissive terms as "a trace." Refinement of analytical techniques in recent decades has allowed much more specific investigation and definition of the role of many trace elements in biological systems.

Chromium, with an atomic number of 24 and a mass of 52.01, belongs to the first series of the transition elements. Its position in subgroup VIB of the periodic system (Figure 1) is surrounded by three elements with known biological function: vanadium, manganese, and molybdenum.[1] Chromium can occur in every one of the oxidation states from -2 to $+6$, but only the ground state 0, $+2$, $+3$, and $+6$ are common. Hexavalent chromium has been considered a toxic metal for many years,[2] perhaps because of the poisonous effects of high exposures to industrial bichromates.[3] Trivalent chromium, on the other hand, is believed to be less toxic. That chromium ions might have biological activity was first suggested in 1954 by Curran,[4] who showed important effects on intermediary metabolism of fats. Schwarz and Mertz[5] in 1959 described trivalent chromium as a factor modifying glucose tolerance. The significance now attached to the biological importance of chromium, not only in the area of adverse effects, but also in the realm of essential function, greatly illustrates Gabriel Bertand's[6] model of the biological dose response to trace elements (Figure 2). This model can be summarized by stating that the biological effect is a function of the element's concentration in the organism. Very low concentrations of an essential element are incompatible with life. With increasing concentrations, specific functions increase to reach a plateau that can be maintained through a range of the element's concentration. Beyond this concentration range, further increases result in deleterious effects—chronic, then acute, toxicity and death. One of the most important objectives of nutritional trace element research is to define those tissue concentrations and those environmental exposures that are compatible with optimal function. The fact that a plateau is maintained throughout a range of tissue concentrations, dietary intakes, and environmental exposures must take into account any homeostatic regulation. It follows from this dose-response concept that trace elements of critical essentiality are inherently toxic and will become so if intake or exposure overcomes normal homeostatic regulation in the organism. This toxicity is not only a function of dosage but also the chemical form in which an element is bound. Selenium and chromium, both now recognized as essential, were first known as toxic. Recent research suggests that even arsenic might be essential.[7]

A proper appraisal of the toxic effects of prolonged or brief exposures to chemical elements and of the possible functions of these elements in human nutrition and metabolism, requires information concerning their normal patterns of absorption and excretion and their occurrence and distribution in various tissues in the body. This chapter, therefore, will now deal with biological and analytical aspects of chromium.

II. BIOLOGICAL ASPECTS

A. Occurrence

The normal body exposure to chromium occurs in food and water. Chromium is ubiquitous in the universe, ranking sixth on the earth's crust, fifteenth in sea water,

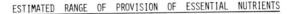

FIGURE 1. Periodic table of elements.

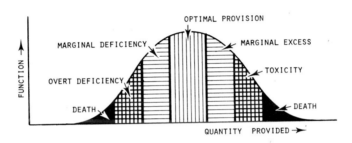

FIGURE 2. Biological dose response to trace elements—after Bertrand.

and fifteenth in the body of man.[1] In crust and sea water, its average concentration is 200 ppm and 1 to 2.5 ppb, respectively. Despite this high concentration in soil, plants are capable of resisting excessive accumulation of the metal. Chromium rich foods are brown sugar, animal fats, and butter. The concentration in natural water supplies is below 10 ppb; however, concentrations of 35 ppb in municipal drinking water supplies have been recorded. The daily intake in man has been estimated at 60 μg (30 to 100 μg) with 10 μg contributed to by water.[8]

B. Absorption

Inorganic trivalent chromium salts are poorly absorbed by man and animals. Chromates are better absorbed, but the preferred valency state for chromium is 3+, and it is likely that any chromates present in the diet are reduced in the gastrointestinal tract from hexa- to trivalency. Using Cr^{51} labeled chromic chloride, Doisy et al.[9] observed that only 0.69% of an oral dose was absorbed by normal human subjects. This agreed with the previously recorded figure of 0.5%.[10] Insulin-requiring diabetics appear to absorb two to four times more chromium than do normal subjects.[9]

C. Tissue Levels

Many of the earlier analytical methods were not sensitive enough to detect chromium in animal tissues. Since chromium was not regularly found, it was not thought to be of biological significance. Data accumulating since 1948 leave no doubt that chromium is a constituent of organic matter. There is a great variation in chromium content between different species. The differences within one species, man, are equally great. Chromium concentrations of 1.7 to 19.5 ppb have been measured in heart muscles[11] while brain tissue has been recorded as 500 to 8000 ppm;[12] this is approximately 40,000 times heart concentrations on a wet weight basis.

Direct elemental analysis on tissue samples of such organs as liver, kidney, muscle, or skin may be useful during research investigations but are seldom practicable in routine patient care. Analysis of hair[13,14] and nail samples may give valuable information of a tendency to element deficiency or excess on a group basis, but are unlikely to be useful for an individual. The majority of analyses will, therefore, be carried out on blood serum and urine.

D. Serum Values

Chromium concentrations in the blood have been reported by many workers.[15–21] Most values are between 20 and 50 ppb, however, the spread of mean values is enormous. They vary between 0.14 and 782 µg Cr mℓ^{-1} serum, with more credibility attached to the lower values. The quality of the sampling and sample preparation are decisive for accurate determinations. There has been a trend in the reported data of serum chromium to follow the limits of analytical detectability available to the investigators involved with the various studies.[20] The variations illustrated by the range of figures available for serum chromium indicate that great care must be taken in this field to eliminate, as far as possible, sampling errors,[22] wrong sampling handling,[23–25] poor analysis, and sometimes poor choice of technique.

Chromium may exist in a variety of forms in the circulation. Chromates are bound largely to red blood cells. Serum chromium exists mainly as Cr^{3+} and bound to plasma proteins, most is found in the β globulin fraction attached to transferrin (siderophilin).[26] When excessive amounts are present, other fractions including albumin become more important. It must be realized that plasma chromium, particularly in the fasting state, is not in equilibrium with the body stores and may not, therefore, adequately reflect tissue loads. It is likely that changes in blood levels after oral loads may be a useful tool in assessing Cr metabolic activities.[27]

E. Excretion

Orally absorbed or injected chromium appears to be excreted mainly by the kidneys; approximately 80% of an injected dose has been recovered in the urine. The daily urinary excretion of chromium varies between 3 and 50 µg/24 hr.[28–30] It is not known how much of this is inorganic chromium or the quantity emanating from metabolic degradation of carbohydrate influencing factors.[31] Much of the data reported in the recent literature has been derived from direct determinations on urine using flameless atomic absorption spectrometry. One must be skeptical as to the accuracy of many readings because matrix composition of the samples may vary considerably.[32] Dilution may be one solution, but this is impracticable in most cases since the detection limit is soon reached. Various urinary constituents, for example, salt content, make interpretation of random samples extremely difficult. Recent studies have suggested that chromium creatinine ratios may be a useful way of assessing changes in chromium metabolism.[33]

F. Environmental Exposure

Occupational exposure to chromium compounds lead to a variety of clinical problems including dermatitis, penetrating ulcers on the hands and forearms, perforation of the nasal septum, inflammation of the larynx and liver, and an increased incidence of bronchogenic carcinoma. The possible mechanisms involved are extremely diverse from hapten sensitization in dermatitis[34] to concentration in nuclear protein in carcinogenesis.[35] It is, therefore, unlikely that present analytical techniques will be useful in elu-

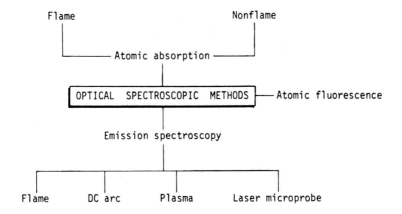

FIGURE 3. Optical spectroscopic methods of trace element analysis.

cidating these detailed interrelationships. The same phenomena as were noted in relation to serum values has been noted in analysis of cement for its chromium content, namely, early reports gave figures of 0 to 0.006%, while recent studies record values of 0 to 0.035%.

III. ANALYTICAL TECHNIQUES

A. Methods

A wide variety of techniques is now available for the analysis of trace elements. Some are relatively simple and have been applied by many workers, others are still, and may always remain for a variety of reasons, strictly limited in application. In general, chromium remains an extremely difficult element on which to obtain reliable data.

1. Optical Spectroscopy

This is perhaps the most widely applied range of procedures in current biological studies. It covers a wide range of procedures (Figure 3). Basic to this group of techniques is the fact that the outer electrons in the molecule can be stimulated to emit or absorb radiation in the optical, or adjacent regions of the electromagnetic spectrum. This is most commonly achieved by heat. If a gas, liquid, or solid sample is heated by a flame, arc, or plasma, then it will emit a diagnostic radiation from which qualitative and quantitative information can be measured. A similarly heated sample can specifically absorb light by a particular series of wavelengths, i.e., atomic absorption, to give measurable data. Flames are the most common heating media used for atomic absorption but nonflame methods are increasing in their application. Atomic fluorescence is an experimental technique which is complementary to atomic absorption and emission and offers certain advantages for some metals, notably zinc.

Flameless atomic absorption spectrophotometry is claimed to have a sensitivity for chromium of 1 to 10 pg. This corresponds to 5 to 50 $\mu\ell$ of a solution containing 0.1 to 1.0 $\mu g/\ell$. This range is only really achievable with aqueous samples which have no complex matrixes. The major difficulty with atomic measurements has been the need for accurate background correction during atomization, especially with serum or urine samples.[32] This also applies to their residual matrix after acid digestion. Apparent chromium values may vary widely depending on the sample preparation technique. Most methods involve preliminary ashing or digestion. Kayne et al.[20] have produced satisfactory data using both fused silica digests and standard addition of inorganic chromium

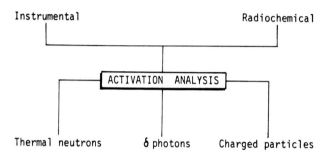

FIGURE 4. Techniques of activation analysis for trace elements.

to materials already oxidized and diluted. These methods allowed routine determination of as little as 0.1 μg/ℓ of chromium in biological fluids. This technique has the advantages of simple sample handling which should decrease contamination and also the ability to deal with a large sample load.

2. Activation Analysis

The sample to be analyzed is irradiated and a small quantity of the natural stable element is converted to a radioactive form (Figure 4). This then decays and in doing so gives off nuclear radiation which can be used to establish the nature and composition of the sample. The most common activation species is thermal neutron emission from a fission nuclear reactor. The methods required to produce energized particles severely limit the availability of these techniques. The other major disadvantage in the method is the length of time needed for sample (30 to 40 days) processing which almost precludes use in an active clinical situation. On the positive side, the method has the advantage that a number of elements may be measured simultaneously, but again, the disadvantage is that the presence of one element may interfere with the measurement of another, both from the point of view of decreasing the specificity of measurement and by increasing the minimum concentration which can be measured. The method has the additional advantage that a reagent blank is not required. Allowing for the problems of application to clinical samples this method[36] as a research tool would seem to have major advantages for chromium measurement in relation to reagent blank and matrix interference problems. Gallorini and Orvini[37] have produced refined methods of sample handling in a variety of biological materials such as bovine liver and pine needles, and produced data of considerable interest. In their analysis, simultaneous measurement of As, Sb, Se, Cr, Cd, and Cu were possible. There is now an increasing interest in this method for Cr determination for the reasons given above.

3. Mass Spectrometry

Mass spectrometry is well-established in medicine and biology as a method for the analysis of chemical composition and this can be applied to trace element determination. The ways in which this method has been developed are shown in Figure 5. The cost, complexity, and precision problems associated with available methods, such as spark source mass spectrometry, have severely limited the application of such methods to trace element analysis in biomedical sciences. Interest in the method, however, has resulted in development of a variety of related techniques such as glow discharge. Isotope dilution and metal chelate production are procedures involving treatment of the sample prior to mass spectrometry, and isotope dilution in particular is an extremely accurate technique for the determination of most biologically important trace elements with the exception of fluorine, cobalt, and manganese. Figures are available for chro-

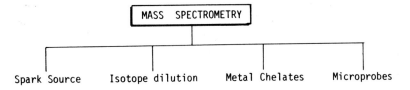

FIGURE 5. Applications of mass spectrometry to trace element analysis.

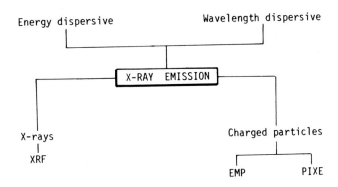

FIGURE 6. X-ray emission techniques in trace element analysis.

mium in human urine and brewer's yeast.[38] These studies indicated that urinary chromium excretion appeared to be an order of magnitude lower than previously reported values. They also supported the concept that deuterium background correction was inadequate for accurate determination of chromium in urine.[32] Also, the data suggested that the method was likely to be useful in the determination of chromium concentrations of below parts per billion in difficult matrices and that the results could be precise and accurate. It must be appreciated that the methods require considerable experience and specially clean laboratories to allow its full potential to be realized.

4. X-Ray Emission Techniques

The variety of techniques is illustrated in Figure 6. Clinicians are already conversant with conventional X-rays. For analytical purposes, X-rays are similarly produced by bombarding a target with electrons and the resultant X-ray spectrum produced is diagnostic of the target composition and so analytical values may be obtained. The two techniques which have been most widely applied to biological samples are proton induced X-ray emission (PIXE) and electron microprobes (EMP). The availability of accelerators with sufficient energy for PIXE, i.e., 1 to 3 MeV, is limited and so data is scant in this area. The other factors limiting its application are the necessary nuclear equipment to detect characteristic X-rays, and the applicability to only these tissue samples (<1 mg cm^2). Its advantages are multielement analysis, nondestructive nature, and high sensitivity (10^{-6} to 10^{-7} g/g). Also, samples (30 to 300 µg) are sufficient for reliable analysis and this brings the method into the realm of needle biopsy specimens. Reports are available on a variety of trace elements in liver biopsies, iron, copper, zinc, manganese, selenium, and molybdenum. No data is available for chromium by this technique.

The EMP, because of the fact that the electron beam is focused, allows analysis of the spatial variation within any given sample. The method just qualifies as a trace element technique since its limit of detection is of the order of 100 ppm or 100 µg/g.

Nevertheless, the ability to localize within tissues particular areas of element concentration may prove extremely valuable for certain clinical situations. In common with all X-ray emission techniques, the interaction of the radiation with the sample means that standardization and sample preparation are extremely critical if quantitative analysis is contemplated.

B. Choice of Analytical Method

In aiming at a choice of method from a wide variety of established and newer developments, the investigator must look for the most suitable method for a particular situation. The final choice is likely to be a compromise since no one procedure is likely to satisfy all criteria. Hislop[39] has suggested that in making this choice it is necessary to consider the following factors.

1. Accuracy and Precision

It is important to distinguish between these two characteristics of analysis. Accuracy is the ability to obtain results which are close to the "true" value, whereas, precision is the reproducibility or variability of the results. Accuracy is, therefore, important where absolute values are required, as in deciding whether a particular element has reached a specific concentration which might indicate toxicity of deficiency. If, on the other hand, it is wished to follow the effect of some procedure, such as elemental dose response, on element concentrations, the ability to measure small changes with confidence is likely to be of much more importance than the accuracy of absolute levels. The important factors in determining accuracy include the specificity of the method, the dependence of the method on the physical and chemical form of the element, and the skill of the analyst. The methods thought to give consistently the most accurate figures are radiochemical activation analysis and isotope dilution mass spectrometry. It has already been noted that these methods are unlikely to be widely available and so limit clinical application severely. They are used currently much more as reference procedures rather than routine tools.

2. Limit of Detection

The limit of detection is the minimum quantity or concentration of a particular substance which can be distinguished from background signals. This background may be produced by interference from other substances, technical considerations, and reagent interactions. The limit, therefore, is determined by the technique employed, apparatus, and laboratory facilities and so comparability of data must be inspected with caution. There is interaction between sample size and experimental limits since small samples with even the best limits of detection may produce poor definitions in respect of concentration limits. Provided there is no limit on sample availability, a situation unlikely in biological situations, the larger the sample and a technique with lower absolute limits of detection may be a more satisfactory approach. The limit of detection may be increased by preanalysis concentration methods. It must be remembered that this will increase potential error or contamination.

3. Elemental Coverage

In relation to presently available techniques, chromium is amenable to analysis by optical, atomic absorption and emission spectrometry, X-ray fluorescence or particle excitation, mass spectrometry by isotope dilution or spark source, and activation analysis of radioactive or γ detection.[40-42] Electrochemical and spin resonance are not of value for this element.

4. Determination of Chemical Form

It is perhaps most usual to use conventional inorganic procedures in association with biochemical separation techniques in order to establish the chemical environment of specific elements.[43] This is because only a limited number of methods are available to analyze this situation directly, namely, spin resonance, magnetic resonance, and Raman spectroscopy. None of these is available for chromium analysis. The importance of chemical form is often of considerable interest in biological materials and certainly in relation to chromium interactions, such relationships are important in various aspects of medical importance. It was thought, for example, that allergy to chromium was limited to hexavalent forms and that allergy to the trivalent form never occurred. This has now been shown to be untrue.

5. Spatial Distribution of Elemental Composition

In trace element analysis, it may be necessary to determine how the concentration of an element varies within a sample. A commonly used biopsy specimen in human elemental research is hair. It has been shown by analyzing 1 cm of hair that important variations in the concentration of Cu, Fe, Co, Ni, Cr, and Mn can occur. By using electron probe techniques such analysis can be taken to variations across the diameter of a hair or within a single cell.

6. Practical Considerations

It is also important to consider the availability of the techniques, the cost, the time taken for analysis, and the sample form as well as the characteristics of the method and information desired. Taking these factors into consideration, it is perhaps not unsurprising that atomic absorption spectrometry is most commonly employed in current clinical chemistry. The technique can be provided in most laboratories, the capital and revenue costs are not excessive, samples can be quickly analyzed in useful numbers, and the technique is perhaps best adapted to liquid samples such as are routine in clinical practice.[44]

C. Future Developments

Among the common determinations carried out in clinical chemistry are those for the elements calcium, chloride, iron, potassium, and sodium. These estimations are carried out mostly in serum and urine. It is arguable whether these elemental analyses are asked for because they can be done, or whether suitable methods have been developed to cope with clinical demand. Requests for analysis of other elements including chromium are much less common, but are likely to be the basis of great advances in our fundamental knowledge of the normal and abnormal metabolism of their elements. The developments required in this field involve not only the widening of understanding of sample materials (other than blood serum and urine, such as hair), but also major improvements in accuracy, precision, and speed of analysis.[45] Chromium analysis must still be regarded as slow, imprecise, and inaccurate in the majority of situations. As an element, it has a long way to go to join rapid or even slow accurate determinations of such substances as sodium or copper.

The main methods of determination used by clinical chemists at present are spectrophotometry (calcium, iron), flame emission spectrometry (potassium sodium), potentiometry (calcium), volumetry (chlorine), and atomic absorption. An estimate of interest in this field of analysis is given recently as atomic absorption 64, potentiometry 23, spectrophotometry 12, neutron activation 8, emission spectrometry 5, voltametry 4, others 9. Atomic absorption, as has been seen, is fast, wide ranging, and sensitive to

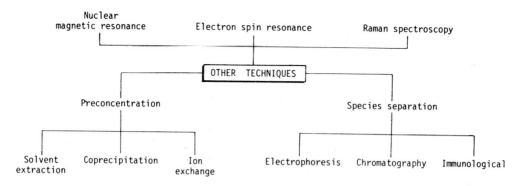

FIGURE 7. Techniques in trace element analysis which involve sample preparation as well as analytical methods.

ng-pg quantities, but is subject to interference in biological matrices; its precision may be poor and may, therefore, lead to data difficult of interpretation. Clearly, methodology will determine to a large extent improved understanding of the role of chromium in the biomedical sciences in the next few years.[46] It is likely that progress will be made by a combination of research into sample processing and newer analytical techniques (Figure 7). At the moment these areas are only beginning to reveal their potential and no useful data is available relating to chromium in clinical practice.

REFERENCES

1. **Mertz, W.**, Chromium occurrence and function in biological systems, *Physiol. Rev.*, 49, 163, 1969.
2. **Akatsuka, K. and Fairhall, L. T.**, The toxicology of chromium, *J. Ind. Hyg. Toxicol.*, 16, 1, 1934.
3. **Brinton, H. P., Frasier, E. S., and Koven, A. L.**, Morbidity and mortality experience among chromate workers, *Public Health Rep.*, 67, 835, 1952.
4. **Curran, G. L.**, Effect of certain transition group elements on hepatic synthesis of cholesterol in the rat, *J. Biol. Chem.*, 210, 765, 1954.
5. **Schwarz, K. and Mertz, W.**, Chromium (III) and the glucose tolerance factor, *Arch. Biochem. Biophys.*, 85, 292, 1959.
6. **Bertrand, G.**, On the role of trace substances in agriculture, *8th. Int. Congr. Appl. Chem.*, New York, 28, 30, 1912.
7. **Anke, M., Grun, M., Partschefeld, M., Groppel, B., and Hennig, A.**, Essentiality and function of arsenic, in *Trace Element Metabolism in Man and Animals—3*, Kirchgessner, K., Ed., Institut für Ernährungsphysiologie der Technischen Universität München, Freising-Weihenstephan, West Germany, 1978, 248.
8. **Hammond, P. B. and Beliles, R. P.**, Metals, in *Toxicology: The Basic Science of Poisons*, 2nd. ed., Doull, J., Klaassen, C. D., and Amdur, M. O., Eds., Macmillan, New York, 1980, 409.
9. **Doisy, R. J., Streeten, D. H. P., Freiberg, J. M., and Achneider, A. J.**, Chromium metabolism in man and biochemical effects, in *Trace Elements in Human Health and Disease—II*, Prasad, A. S., Ed., Academic Press, New York, 1976, 79.
10. **Donaldson, R. M. and Barreras, R. F.**, Intestinal absorption of trace quantities of chromium, *J. Lab. Clin. Med.*, 68, 484, 1966.
11. **Webster, P. O.**, Concentration of twenty four trace elements in human heart tissue, determined by neutron activation analysis, *Chem. Abstr.*, 61, 13683, 1964.
12. **Voinar, A. O.**, Amount of microelements in nuclei of nerve cells measured by emission spectra, *Biokhimiya*, 18, 29, 1953.
13. **Hambidge, K. M. and Droegemueller, W.**, Changes in plasma and hair concentrations of zinc, copper, chromium and manganese during pregnancy, *Obstet. Gynecol.*, 44, 666, 1974.
14. **Klevay, L. M.**, Hair as a biopsy material, *Am. J. Clin. Nutr.*, 23, 284, 1970.

15. **Imbus, H. R., Cholak, J., Miller, L. H., and Sterling, Y.**, Boron, cadmium, chromium and nickel in blood and urine, *Arch. Environ. Health,* 6, 286, 1963.
16. **Davidson, I. W. F. and Secrest, W. L.**, Determination of chromium in biological materials by atomic absorption spectrometry using a graphite furnace atomizer, *Anal. Chem.,* 44, 1808, 1972.
17. **Grafflage, B., Buttgereit, G., Kübler, W., and Mertens, H. M.**, Die Messung der Spurenelemente Chrom und Mangan im Serum mittels flammenloser Atom-absorption, *Z. Klin. Chem. Klin. Biochem.,* 1974.
18. **Li, R. T. and Hercules, D. M.**, Determination of chromium in biological samples using chemiluminescence, *Anal. Chem.,* 46, 916, 1974.
19. **Versieck, J., Hoste, J., Barbier, F., Steyaert, H., De Rudder, J., and Michels, H.**, Determination of chromium and cobalt in human serum by neutron activation analysis, *Clin. Chem.,* 24, 303, 1978.
20. **Kayne, F. J., Komar, G., Laboda, H., and Vanderlinde, R. E.**, Atomic absorption spectrophotometry of chromium in serum and urine with a modified Perkin-Elmer 603 atomic absorption spectrophotometer, *Clin. Chem.,* 24, 2151, 1978.
21. **Veillon, C., Wolf, W. R., and Guthrie, B. E.**, Determination of chromium in biological materials by stable isotope dilution, *Anal. Chem.,* 51, 1022, 1979.
22. **Versieck, J., Speecke, A., Hoste, J., and Barbier, F.**, Trace element contamination in biopsies of the liver, *Clin. Chem.,* 19, 472, 1973.
23. **Wolf, W. and Green, E.**, Preparation of biological materials for chromium analysis, in Accuracy in Trace Analysis Sampling, Sample Handling and Analysis, (7th Materials Res. Symp. Gaithersburg, MD., 1974) U.S. Government Printing Office, Washington, D.C. 1976, 605.
24. **Behne, D.**, Problems of sampling and sample preparation for trace element analysis in the health sciences, in *Trace Element Analytical Chemistry in Medicine and Biology,* Brätter, P. and Schramel, P., Eds., Walter de Gruyter & Co., Berlin, 1980, 769.
25. **Moody, J. R. and Lindstrom, R. M.**, Selection and cleaning of plastic containers for storage of trace element samples, *Anal. Chem.,* 49, 2264, 1977.
26. **Jones, H. D. C. and Perkins, D. J.**, Metal-ion binding of human transferrin, *Biochim. Biophys. Acta,* 100, 122, 1965.
27. **Mali, J. W. H., Van Kooten, W. J., and Van Neer, F. C. J.**, Some aspects of the behaviour of chromium compounds in the skin, *J. Invest. Dermatol.,* 41, 111, 1963.
28. **Schroeder, H. A.**, The biological trace elements or peripatetics through the Periodic Table, *J. Chronic Dis.,* 18, 217, 1965.
29. **Gürson, C. T. and Saner, G.**, Urinary chromium excretion, diurnal changes and relationship to creatinine excretion in healthy and sick individuals of different ages, *Am. J. Clin. Nutr.,* 31, 1162, 1978.
30. **Davidson, I. W. F., Burt, R. L., and Parker, J. C.**, Renal excretion of trace elements: chromium and copper, *Proc. Soc. Exp. Biol. Med.,* 147, 720, 1974.
31. **Schroeder, H. A., Balassa, J. J., and Tipton, I. H.**, Abnormal trace metals in man—chromium, *J. Chronic Dis.,* 15, 941, 1962.
32. **Guthrie, B. E., Wolf, W. R., and Veillon, C.**, Background correction and related problems in the determination of chromium in urine by graphite furnace atomic absorption spectrometry, *Anal. Chem.,* 50, 1900, 1978.
33. **Saner, G.**, Urinary chromium/creatinine ratio in the assessment of chromium nutritional state, in *Trace Element Analytical Chemistry in Medicine and Biology,* Brätter, P. and Schramel, P., Eds., Walter de Gruyter & Co., Berlin, 1980, 159.
34. **Fregert, S. and Rorsman, H.**, Allergy to trivalent chromium, *Arch. Dermatol.,* 90, 4, 1964.
35. **Herrmann, H. and Speck, L. B.**, Interaction of chromate with nucleic acids in tissues, *Science,* 119, 221, 1964.
36. **Draskovic, R. J., Jacimovic, L. J., and Draskovic, R. S.**, Some applications of activation analysis in medicine, in *Trace Element Analytical Chemistry in Medicine and Biology,* Brätter, P. and Schramel, P., Eds., Walter de Gruyter & Co., Berlin, 1980, 173.
37. **Gallorini, M. and Orvini, E.**, The role of radiochemical neutron activation analysis in certifying selected trace elements content in biological related matrices, in *Trace Element Analytical Chemistry in Medicine and Biology,* Brätter, P. and Schramel, P., Eds., Walter de Gruyter & Co., Berlin, 1980, 675.
38. **Harrison, W. W. and Bentz, B. L.**, Trace element analysis by glow discharge mass spectrometry, in *Trace Element Analytical Chemistry in Medicine and Biology,* Brätter, P. and Schramel, P., Eds., Walter de Gruyter & Co., Berlin, 1980, 285.
39. **Hislop, J. S.**, Choice of the analytical method, in *Trace Element Analytical Chemistry in Medicine and Biology,* Brätter, P. and Schramel, P., Eds., Walter de Gruyter & Co., Berlin, 1980, 747.
40. **Evenson, M. A. and Carmack, G. D.**, Review of trace element analytical techniques, *Anal. Chem.,* 51, 35R, 1979.

41. **Morrison, G. H.**, Trace element analysis in human serum, *Crit. Rev. Anal. Chem.*, 8, 287, 1979.
42. **Fell, G. S., Shenkin, A., and Halls, D. J.**, Trace element analysis as a diagnostic tool in clinical medicine, in *Trace Element Analytical Chemistry in Medicine and Biology*, Brätter, P. and Schramel, P., Eds., Walter de Gruyter & Co., Berlin, 1980, 217.
43. **Feldman, F. J.**, The state of chromium in biological materials, *Fed. Proc. Fed. Am. Soc. Exp. Biol.*, 27, 482, 1968.
44. **Feinendegen, L. E. and Kasperek, K.**, Medical aspects of trace element research, in *Trace Element Analytical Chemistry in Medicine and Biology*, Brätter, P. and Schramel, P., Eds., Walter de Gruyter & Co., Berlin, 1980, 1.
45. **Evenson, M. A.**, Requirements for clinical instruments in trace element analysis, *Anal. Chem.*, 51, 1411 A, 1979.
46. **Bowen, H. J. M.**, The requirements of medicine for trace element analysis in the 21st century, in *Trace Element Analytical Chemistry in Medicine and Biology*, Brätter, P. and Schramel P., Eds., Walter de Gruyter & Co., Berlin, 1980, 783.

Chapter 2

THE CARCINOGENICITY OF CHROMIUM COMPOUNDS IN MAN AND ANIMALS

Sverre Langård

TABLE OF CONTENTS

I. Introduction 14
   A. Historical Perspective 14
   B. Possible Carcinogenic Hazards of Chromates 14
      1. General Environment 14
      2. Occupational Exposure 14
      3. Exposure through Foodstuffs 15
      4. Chromium Compounds in Treatment of Human Diseases 15

II. Epidemiologic Studies in Humans 15
   A. Epidemiologic Considerations 15
      1. Minimum Initiation Dose 15
      2. Latent Period 15
      3. Observation Time 17
      4. Relationship between Dose and Effect 17
      5. Other Dose Estimates 18
   B. Reviews on the Carcinogenic and Mutagenic Effects of Chromium 18
   C. Production of Chromium Compounds 18
      1. Monochromates and Dichromates 18
      2. Ferrochromium Production 20
      3. Production of Chromium Pigments 20
   D. Users of Chromium Compounds 21
      1. Chromium Pigments 21
      2. Chromium-Plating Industries 22
      3. Stainless Steel Welding 22

III. Experimental Studies in Animals 23
   A. The Animal Model 23
   B. Evaluation of the Results 24

IV. Conclusion and Need for Further Research 25
   A. Importance of Water Solubility 25
   B. Dose-Response Relationship 25
   C. Research Needs 26

References 26

I. INTRODUCTION

A. Historical Perspective

It has been 50 years since Lehmann[1] in 1932 suggested that occupational exposure to chromates was associated with increased risk of developing lung cancer. His suggestion was based on the occurrence of two cases of lung cancer among several hundred chromate workers in Ludwigshafen, West Germany; cases which were observed by Koelsch in 1912. As early as in 1890, Newman[2] reported one case of adenocarcinoma in the left inferior turbinated body of the nose in a chrome pigment worker who also had perforation of the nasal septum. Reports by Alwens et al.[3] and Pfeil[4] further documented the association between inhalation exposure to dust in dichromate manufacturing industry and development of lung cancer in the workers. These reports made the German health authorities accept lung cancer in chromate producing workers as an occupational disease in 1936.[5]

As documented by MacLure and MacMahon,[6] the history of chromates as carcinogens represents one of the depressing examples where knowledge, which has been available in the literature for many decades, has not led to the desired prevention of preventable cancer in man. Even in 1981[7] and 1982,[8] 45 years after it was acknowledged that chromates were carcinogenic in man, studies appear on the high incidence of lung cancer in workers exposed to chromates.

B. Possible Carcinogenic Hazards of Chromates

Chromium compounds are present in human diets,[9,10] in environmental air[11] even in remote places,[12] in natural waters,[13] in seawater,[14] and in plants.[15] Hence, chromium compounds may be considered to be ubiquitous. Consequently, if most of the chromium compounds which occur in the diet and the environment were carcinogens, these compounds might constitute a cancer hazard to humans anywhere in the world. However, experience shows that the effects on humans are less serious than this grim outlook might suggest.

1. General Environment

Although it has been documented that the general environment may be heavily contaminated by effluents from chromium manufacturing factories[16] and it has been suggested that this may impose a cancer hazard to humans,[17] no hazard was documented by Axelsson and Rylander[18] in a study which was designed to elucidate this particular question. They studied the cancer incidence in residents in the neighborhood of a ferrochromium smelter in Sweden. The dilution of the possible carcinogenic chromium compounds in the effluents from this ferrochromium smelter is considerable. Therefore, such a nonpositive study[18] should be interpreted with caution. Still the possibility should not be excluded that chromium compounds in the general environment in certain regions may make a slight contribution to the incidence of cancer experienced in the local residents.

No studies have been carried out on possible carcinogenic hazards of chromates in municipal water. The presence of such a hazard seems unlikely.

2. Occupational Exposure

Most of the available knowledge on the carcinogenic effects of chromium compounds has been generated from studies on occupationally exposed groups. This subject will be dealt with in Section II of this chapter.

3. Exposure through Foodstuffs

Certain chromium compounds are considered to be essential to man[19] and are required in the diet. A wrong composition or excessive amount of these compounds in the diet, therefore, could influence the occurrence of cancer in the gastrointestinal tract. However, no evidence has been presented supporting this possibility.

4. Chromium Compounds in Treatment of Human Diseases

Alloys of chromium have been in use as joint replacement prostheses in humans for many years.[20] Heath et al.[21] injected rats intramuscularly with wear particles from prostheses made in cobaltchromium alloys and found a number of sarcomas at the injection sites. This result made him suggest that a carcinogenic hazard might be present for humans using these prostheses. However, this study has been criticized[22] because too many far-reaching conclusions were drawn on the basis of animal experiments and at the present time no conclusive evidence[23] has been presented supporting these suggestions.[21]

Chromium trioxide (CrO_3) has been in use for many years as a cauterizing substance, mainly to stop nose bleeding. The possibility of a carcinogenic hazard associated with this use of CrO_3 has not been evaluated. Some evidence has, however, been presented that exposure to aerosols deriving from CrO_3 may be associated with increased lung cancer risk,[24-26] and that CrO_3 acts as a mutagen.[27] Therefore, this compound should be considered as a carcinogen.[20]

Radiochromates are widely used to evaluate the lifespan of red blood cells,[28] lymphocytes,[29] and platelets,[30] and have also been used for labeling human liver cells.[31] Recently, radiochromates have also found application in early diagnosis of gastrointestinal bleeding.[32] No carcinogenic hazard has been reported associated with the use or application of these trace amounts of chromates.

II. EPIDEMIOLOGIC STUDIES IN HUMANS

A. Epidemiologic Considerations

1. Minimum Initiation Dose

Epidemiologic studies on the carcinogenicity of chromium compounds have been carried out during the last 35 years. During recent years, increasing emphasis has been put on the characterization of level of exposure and duration of exposure. These methodological changes have to be taken into account when reviewing literature covering such a long period.

Also in the last few years, more prominence has been given to the question "what is the minimum exposure time necessary to initiate tumor development" and to the "latent period" of cancer development. In this context the *sufficient "dose"* or *minimum initiation "dose"* may be *defined* as "the minimum time of exposure which is needed to increase the cancer incidence sufficiently in the population at risk so that the increase can be demonstrated after the latent period has elapsed, when using the presently available methods" (Figures 1 and 2).

2. Latent Period

This term is *defined* as the time period elapsing between the moment when the minimum initiation dose has been reached and the moment when the first agent-induced tumor has developed to a size that permits recognition and diagnosis.

Since the tumor is not "sleeping" during this latent period, the term in itself is to

16 Chromium: Metabolism and Toxicity

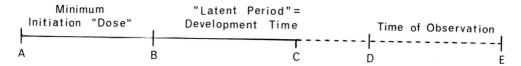

FIGURE 1. The figure indicates the situation for one subject who meets the criteria for being included in an epidemiologic study. The letter A indicates the first day of exposure (employment); C is the last day of exposure; E is the last day of observation for the whole study-group. The interval between the letters A and B illustrates the minimum exposure time assumed to be sufficient to increase the tumor incidence to a measurable magnitude when using the available epidemiologic methods. This interval equals the usual exposure time criterion for inclusion in the study population. The time interval between the letters B and D represents the experienced tumor development time. This interval should, in contrast to the common practice, not be included in the calculation of expected number of cancer cases. Only the interval between the letters D and E should be included in the calculation of expected figures for each individual subject in the study population.

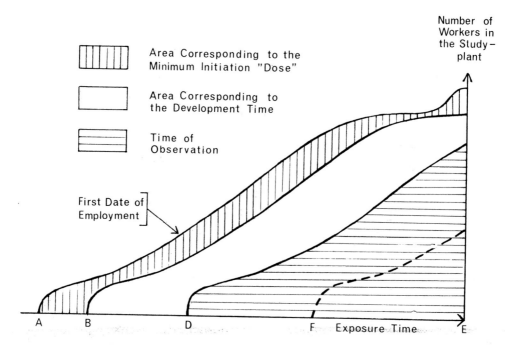

FIGURE 2. Estimation of expected number of cancer cases in a cohort study. The figure illustrates the total number of present and previously exposed workers in a plant or industrial setting who are recruitment subjects for a cohort study. The letters A (first day of exposure), B (end of minimum initiation "dose"), and D (end of latent period) correspond to the letters in Figure 1. Only those workers who have been employed for a longer time period than that which is defined as the minimum initiation dose should be included in the study cohort. Those who meet this criterion serve for estimation of expected number of cases, but only from the day when the development time has elapsed (D). Recently employed workers should also be excluded from the calculation of expected figures; this concerns all workers who have been employed more recently than the sum of the minimum initiation dose and the development time. The letter F corresponds to this date. If 5 years has been chosen for the minimum initiation "dose" and 10 years as development time, all workers engaged during the last 15 years should be excluded from the calculation of expected figures.

some extent misleading. Therefore, the term *development time* should be used instead.

In assessing epidemiologic data, it is of great importance to consider these different terms and to try to extract whether these terms have been defined and taken into consideration in the method section of the papers. When minimum initiation dose and latent period have not been taken into consideration when defining the relevant study group in cohort-studies and case-referent studies, the results may include a considerable "dilution" factor, making otherwise statistically significant results nonsignificant.[33] Some authors also apply the term "latent period" to the time between the first day of exposure and the moment when the first agent-induced tumor can develop, i.e., the sum of minimum initiation dose and latent period as defined above. In the present review, the different epidemiologic studies on the association between exposure to chromium compounds and cancer will be discussed as far as possible in the light of the differentiation between these terms. These definitions are illustrated in Figures 1 and 2.

3. Observation Time

When estimating the expected number of deaths or cancers in cohort studies, different methods have been applied. Some authors compute the expected figures from each worker's first day of employment to the end of the observation period, including workers who have been first employed during the last 10 to 15 years. This way of calculating the expected number includes both the time for the minimum initiation dose and the latent period. Since the exposed worker has no increased risk of developing cancer during this period, as compared with the reference group, this way of calculating the expected figures implies a considerable dilution of the observed/expected figures, which may be critical for a significant result when the hazard is not very great. Therefore, the observation period should start at the end of the latent period. Consequently, recently employed workers who have been employed for shorter periods than the sum of the time for minimum initiation dose and the latent period, should also be excluded from the study population.

4. Relationship between Dose and Effect

The term *dose* has been *defined* in different ways in industrial toxicology. Many of the toxicologic definitions are not applicable in epidemiologic research. In epidemiology, one is often forced to restrict the dose definition to the following: the amount of a given substance inhaled by—or otherwise having influence on—a subject during a defined time period. This definition requires that the concentration of the substance in the working atmosphere has been determined continuously. In many cases the exposure level has not been determined, therefore, the *duration of exposure* is left as the only simple criterion for dose estimation. Ideally, in epidemiologic studies, the dose should be defined as the amount of a given substance which is taken up in the exposed body during a given time. Since this ideal dose definition requires that a number of additional parameters have to be determined, such as the particle size distribution of an inhaled aerosol, the solubility of the aerosol, and the deposition and retention in the lung, such a definition is not versatile in epidemiology.

Even the simplest definition of dose appears to be difficult to fulfill in epidemiologic studies. A number of the studies on chromate cancer have included measurements of atmospheric concentrations of the agent(s) concerned, but in most cases the measurements have only been carried out during a short time period at the end of the observation period,[34-39] therefore the presented measurements do not necessarily reflect the true exposure level at the time when the cancers were initiated. Consequently, even if the measurements of atmospheric concentrations have been carried out adequately in the different studies, it is apparent that the presented measurements do not provide sufficient information on the dose estimation in order to justify conclusions on a quantitative dose-effect relationship.

5. Other Dose Estimates

Since measurements at the site do not yet give adequate dose estimates, the studies on chromate-induced cancer should be judged by their use of other means to characterize exposure. In the absence of good atmospheric measurements, an accurate job characterization and occupational history for each worker remains a very important source of information for the determination of magnitude of exposure. Therefore, the quality of chromate cancer studies should be judged by the quality of the way the authors have been able to differentiate between the various exposure-characterized subgroups within the cohort.

B. Reviews on the Carcinogenic and Mutagenic Effects of Chromium

Baetjer[40] was one of the first to review the literature presented up to 1950 on the occurrence of cancer in chromate-exposed workers. She covered mainly German and American literature. In 1953, Gafafer[41] presented an extensive review of the whole literature concerning the carcinogenic effects of chromium compounds in man and animals. In 1973, the National Institute for Occupational Safety and Health (NIOSH)[42] covered the literature on the possible carcinogenic effects of chromium trioxide and in 1975, NIOSH[43] presented an extensive review of the toxicity and carcinogenicity of chromates. In this NIOSH document[43] differentiation was made between what was called "carcinogenic" and "noncarcinogenic" chromates. The evidence which supports such a differentiation is weak or nonexistent; consquently this conclusion has been criticized in more recent reviews.[20,44,45] Enterline[46] in 1974 reviewed the epidemiologic studies on chromate-induced cancer. Langård and Norseth[47] published a short review on the carcinogenicity and toxicity of chromium compounds, while one of the most extensive reviews during the recent years has been presented by the U.S. National Academy of Science.[48] Recently, Norseth[49] surveyed the latest literature on the topic. An excellent review has been presented by Hayes[50] and recently another excellent review on chromium toxicity and carcinogenicity was presented by the International Agency for Research on Cancer (IARC).[51]

The mutagenicity of chromium compounds has recently been covered by Léonard and Lawreys.[52]

C. Production of Chromium Compounds

1. Monochromates and Dichromates

Most of the presented epidemiologic studies on the relationship between exposure to chromates and cancer development have been carried out on workers in the chromium chemical industry. This topic has recently been excellently reviewed by Hayes[50] and has also been covered in other recent reviews and surveys.[20,42–49,51] An early review in 1937 by Corozzi[53] should also be mentioned.

Alwens and Jonas[54] reported 20 cases of lung cancer in dichromate-producing workers in a plant near Frankfurt in West Germany. In this study, the time between the start of exposure and the time when cancer was diagnosed varied between 22 and 40 years, with 30 years as the mean. No information was presented on the smoking habits of the workers. Quantitative analysis of the chromium content in the lungs was carried out in some of the cases and high contents were found. However, since the analysis methods were quite different from the methods used today, the results are difficult to compare with recent results.

The results presented in studies in the early 1930s made the German authorities recognize lung cancer (occurring in chromate workers) as occupational cancer in 1936.[5] Based on the observation that exposure to high concentrations of chromate dust may lead to perforation of the nasal septum of the workers, legislation on maximum allowable concentrations of chromates was introduced in the U.S. in 1928.[55] However, no cases of lung cancer were officially accepted as being caused by chromate exposure in the U.S. before 1946.[56] Recognition came as a result of litigation.

Subsequently, Machle and Gregorius[34] carried out a study in six major chromate-producing plants in the U.S. They found that among the 156 deaths in their study group, 46 deaths were due to cancer, of these, 32 were due to lung cancer. According to their presentation, the lung cancer death figures represented an excess of 16 times the expected ratio. Their findings were based upon assessment of proportional mortality; minimum initiation dose and latent period were not taken into consideration, retired workers were not included, and the reference population was not comparable in all relevant respects. Therefore, it is most likely that their findings represent a considerable underestimation of the observed/expected ratio.

Bidstrup[57] and Bidstrup and Case[58] studied the occurrence of lung cancer in dichromate-producing plants in England. In the first paper,[57] only one lung cancer case was recorded, therefore the study was inconclusive. In the follow-up study[58] 12 lung cancer cases were observed, against 3.3 expected. Of the 724 workers included in the first study,[57] 237 had been employed in the chromate industry for more than 15 years. In the follow-up study,[58] no information was given as to how many, or if all, of the 12 lung cancer cases were recruited from this subgroup. If they were all included, only this group could have served as a base for the estimation of the expected number of cases, which would then become lower than the mentioned figure. Inclusion of the rest of the study population when calculating the expected figures leads to a dilution of the results.

The chromate-producing population studied by Bidstrup[57] and Bidstrup and Case[58] was followed up by Alderson et al.[7], who also included those workers employed in subsequent years. This study included 2715 men. A number of improvements in the hygienic conditions in the plant had been introduced in the meantime. They found 116 lung cancer cases in the population, against 48 expected. The relative risk of developing lung cancer had dropped from 3.0, before the modifications of the plant, to 1.8 after the improvements. They also observed two deaths due to nasal cancer.

In a case-referent study on 198 male lung cancer cases in the Johns Hopkins Hospital and on 92 male cases in the Baltimore City Hospital, Baetjer[59] found that 7 and 4 of the cases, respectively, had previously worked in chromate-producing industry, while none of the members of the control series had worked in chromate industry. She used a large number of referents and the relative risk in this study was 28.

Mancuso and Hueper[35] and Mancuso[36] carried out a cancer study in a chromate plant in Ohio, using proportional mortality in their analysis. Six out of nine deaths due to cancer were caused by lung cancer, which gave 17 as observed/expected ratio.

Taylor[60] in 1966, presented an extensive follow up study of males born after 1890 who were employed in one of the U.S. chromate-producing plants between 1937 and 1940. The study covered a period of 24 years and included 1212 workers. Information on deaths in the cohort was derived from Old-Age and Survivers Disability Insurance (OASDI) and the whole U.S. male population was used as reference. Latent period was not subtracted when calculating the expected death figures. These weaknesses in the study design should be taken into account when examining the paper. Taylor found that the age-specific survivorship decreased with increasing employment time. He also demonstrated an increasing lung cancer incidence with increasing duration of work; the observed/expected ratio for lung cancer varying from 4 to 18. Deaths due to other cancers were also in slight excess in the study population.

Ohsaki et al.[61] presented a study on a quite small population of dichromate and chromium trioxide-producing workers in Japan. They observed 14 lung cancer cases in a group of 554 workers. All cases were found within a sub-group of 133 who had worked for more than 10 years in the plant. Using the expected lung cancer rates estimated on the basis of all 554 participants, the excess was more than 50 times the expected figures. All but one of the lung cancer cases occurred in smokers. Four of the cases were small cell anaplastic cancers, while the others were squamous cell cancers. Only few details were given on the epidemiologic methods used in this study.

In a well-designed study, Hayes et al.[62] studied 2101 dichromate-producing workers intially employed between 1945 and 1974, and who had worked in the industry for more than 90 days. The results in this study reflect to a great extent the cancer hazard in the American chromate-producing industry after improvements in the hygienic standards had been introduced in the early 1950s. The population was subdivided into those who were initially employed from 1945 to 1949, from 1950 to 1959, and from 1960 and onwards. Since the workers first employed after 1960 had not been under observation long enough to develop chromate-induced cancer, no cancer cases were observed in this sub-group. In the other two sub-groups the observed/expected ratio varied between 1.8 and 3.4, without any significant difference between the groups. In this study, the highest cancer risk was observed in the group of workers employed at the "wet end" of the production process. Attempts were also made to analyze the interaction with tobacco smoking, but the quality of the records on smoking habits were not good enough to allow analysis.

2. Ferrochromium Production

Pokrovskaya and Shabynina[63] suggested that in ferrochromium workers there was an increased risk for lung cancer and cancer in the gastrointestinal tract. However, too few details were given to allow evaluation of this study.

Two Scandinavian studies[38,39] were carried out at the same time with identical study designs to identify possible cancer hazards in this industry though it was assumed that the workers were mainly exposed trivalent chromium compounds. In both studies, however, hexavalent compounds were identified in the working atmosphere. In one of the studies,[38] a slight excess of lung cancer was demonstrated in workers who had been employed at the ferrochromium furnaces for many years. This excess varied from 2.3 when compared with the total Norwegian male population to 8.5 when using an internal reference population. In this study, it is suggested that hexavalent chromium compounds are the major causative agents for the lung cancer excess. In the Swedish study,[39] no increase in the lung cancer incidence was demonstrated.

Consequently, no conclusive evidence has been presented so far on the presence of a cancer hazard in the ferrochromium-producing industry.

3. Production of Chromium Pigments

The first cancer case ever reported[2] to be related to chromium pigment exposure was seen in a 47-year-old man who had been exposed to chromium pigments.[2] Gross and Kölsch,[64] in 1943,[4] reported eight lung cancer cases in workers in three different chromium pigment-producing plants. These cases were diagnosed at the age of 34, 35, 36, and 36 years of age, and at 42, 45, 45, and 61 years. No details were available on the exposure level in the plants and smoking habits of the workers were not reported. However, nasal perforation was reported in three out of five cases which were presented in detail, indicating that the exposure level may have been quite high. Although the kind of pigment exposure was not reported in detail, zinc chromate had apparently been the most important exposure factor in these five patients who had been exposed to chromium pigments for 5 to 17 years (mean 11.5).

In 1950, Baetjer[40] reviewed the literature on the relationship between exposure to chromates and cancer, and found that 11 cases had been reported on the relationship between exposure to chromium pigments and cancer, including those reported by Gross and Kölsch.[64]

In 1975, the first epidemiologic study on the relationship between chromium pigments and cancer appeared.[37] In a small plant which had employed 133 men from 1948 to 1973, 3 lung cancer cases had occurred in those 24 workers who had worked for more than 3 years before 1973. This study has been criticized because of its small size,[65] but in a follow-up study on the same population to the end of 1981,[8] 3 additional cases of lung cancer had occurred in the preselected group of 24. In neither of these studies has the development time (latent period) been excluded when calculating the "expected" incidence, therefore, the expected figures calculated in the studies are too high.[66] Except for one of the cases, these cancer patients had been exposed exclusively to zinc chromate, indicating that this pigment was the cancer-causing agent in this study. The mean exposure time was 11.5 years and the mean time between initiation of exposure and cancer diagnosis was 18 years. A slight excess of cancer in the gastrointestinal tract in the workers was also observed in this population.[67]

In 1976,[68] an American study was presented including three different plants which mainly produced lead chromate. There was a slight excess of lung cancer in the workers who had been employed for more than 10 years before 1960, and in one of the plants, which produced both lead and zinc chromates, there was a slight increase in gastric cancer in the workers. It is noteworthy that the increased cancer incidence in this study was considerably lower than in the Norwegian study.[37] Although no exact information was presented on the exposure level,[68] it seems likely that the difference between the study results is mainly due to a lower exposure level in the American plants than in the Norwegian one.

In a study in the British chromium pigment industries[69,70] there were 25 lung cancer deaths against 10 expected. Two workers had died of lung cancer in their thirties, and 4 in their forties. In this study too, it appeared difficult to differentiate between the cancer risk associated with exposure to zinc chromate and lead chromate. However, the author suggests that zinc chromate may have been the major exposure factor. Also in this study the relative risk of developing lung cancer was considerably lower than in the Norwegian study.[37]

In a study performed in other European chromium pigment plants presented in 1980,[71] there was also a slight excess of lung cancer in the workers. However, this study has not yet been completed.

D. Users of Chromium Compounds

1. Chromium Pigments

Dalager et al.[72] studied the occurrence of cancer in zinc chromate spray painters and found 21 cancer cases in the respiratory organs against 11.4 expected. This study indicates that a lung cancer hazard is not only present among the producers of chromium pigments, but also in the users of the pigments.

In another American study presented in 1980[73] no excess of lung cancer was seen in a group of automobile spray painters who had been spraying with chromium pigments for more than 5 years before 1955. Therefore, the studies on lung cancer incidence in users of chromium pigments are not consistent. Differences in exposure level and the use of personal protection of the respiratory organs could be responsible for the differences between the study results.

2. Chromium-Plating Industries

Chromium plating is being carried out in a large number of small work places. Hence, it has been quite complicated and tedious to carry out epidemiologic studies in such industries. Chromium trioxide (CrO_3) is the prime agent used in this industry. Since chromium is present in the hexavalent form, one would expect that exposure to chromium trioxide mist would be associated with a cancer hazard in the workers.

However, no epidemiologic study on this topic appeared until 1975, when Royle[25] presented his retrospective mortality study on 1238 platers, among whom 109 had died before June 1972. In this study a matched reference population of similar size consisting of manual workers was used, matched for age, sex, and smoking habits. Consequently, it was critical that the latent period was excluded in the calculation of expected figures. In the study population, 17 lung cancer cases were observed against 10 expected, and 22 cancer cases at other sites among the platers, against 11 in the referents. There were 9 cases of cancer were observed in the gastrointestinal tract in the platers against 4 in the reference population.

In 1975, Waterhouse[74] presented preliminary results of a study in nearly 5000 chromium platers, of whom about half were defined as "chrome bath workers". Within the total group, 49 lung cancer cases were observed against 35 expected. This preliminary result indicates that the population is quite interesting and should be studied further in order to give conclusive results. When subdividing this cohort in different exposure categories and observation periods, it may be possible to determine dose-response relationships within the population.

Okubo[75] presented a cancer incidence study on 889 male Japanese chromium platers, who were followed from April 1970 to September 1976. Since the cohort was based on union membership in April 1970, the nonpositive results presented in this study do not permit any conclusion. Follow-up studies on this population may become very valuable.

Since epidemiologic studies on the relationship between exposure to chromic acid mist and cancer development are few, case histories should also be reported. In 1958 Barbořík et al.[24] reported a lung cancer case in a 35-year-old man who had been a plater for 10 years. Similarly in 1965 Sehnalová and Barbořík[76] reported one case of lung cancer in a 39-year old plater, and in 1977 Michel-Briand and Simonis[26] presented two lung cancer cases in platers aged 41 and 47. Although no final conclusions can be drawn from such case reports, the young age of the cancer patients should be taken into account when considering the possible carcinogenic effects of the water soluble compound chromium trioxide. Therefore, chromic acid should not be considered non-carcinogenic as suggested by NIOSH in 1973.[42]

Documentation of mutagenic effects of chromium trioxide in strains of *Salmonella typhimurium*[77] also support the view that chromic acid should be considered as a human carcinogen.

3. Stainless Steel Welding

Stainless steel is an alloy which contains mainly iron, but also a number of other metals. Steel contains from 10 to 20% metallic chromium and from 0.5 to 20% nickel. The most commonly occurring alloy contains 18% chromium and 8% nickel. The electrodes used contain about the same amount of chromium and nickel as the steel itself. Although it has been known for many years that welding fumes derived from stainless steel welding contain hexavalent chromium compounds,[78] the possible carcinogenic hazard associated with welding has attracted little interest in previous years.

In recently performed studies, the occurrence[79] of hexavalent chromium compounds in the welding fumes and the factors influencing the amount of hexavalent chromium in the fumes[80] have been extensively studied. The amount of hexavalent chromium compounds in the fumes appeared to be much higher when using manual metallic arc (MMA) welding and coated electrodes than when using metal inert-gas (MIG) welding; in both cases stainless steel containing 18% chromium and 8% nickel was welded.[80]

A number of in vitro studies have been carried out, exposing bacteria and cell lines to welding fumes deriving from stainless steel welding. These studies strongly suggest that the welding fumes contain mutagenic components.[81-84] It was also shown that the hexavalent chromium compounds in the fumes are the main contributors to the mutagenic action of the fumes.[82]

Only a few epidemiologic studies have been performed on welders as a separate study entity. In most investigations where welders are studied, the welder group has been considered as a separate subgroup within the study population. In a case-control study on petroleum industry workers, Gottlieb[85] included welders as a separate group and found eight lung cancer cases among the welders against two among the controls. However, the information on welding exposure was too limited to draw conclusions on causal relationship between stainless steel welding and occurrence of cancer in the welder subgroup. Sjögren[86] carried out a retrospective cohort study on 234 welders who had had stainless steel welding as their major task for more than 5 years between 1950 and 1965. He found 3 lung cancer cases against 0.7 expected.

Breslow[87] studied the smoking histories and occupations of 518 confirmed lung cancer patients in a hospital and found that 10 cases had been welders for more than 5 years against 1 of the controls. When sheet metal workers were also included, 14 cases had occurred in workers with these occupations against 2 among the controls. Again, too few details were presented on the kind of welding the cancer patients and controls had been doing.

Milham[88] found a proportional mortality ratio of 119 on lung cancer in a large group of welders who were included in a study on cancer mortality in metal workers. Menck and Henderson[89] demonstrated a standardized mortality ratio (SMR) of 137 for lung cancer in a large mortality survey including 15,300 welders. Finally, Beaumont and Weiss[90] found an SMR of 131 for lung cancer in a group of 3250 welders (53 observed vs. 40.3 expected). When only the person-years at risk after 20 years from first employment were considered, the SMR for lung cancer was 169 (40 observed vs. 23.7 expected). A common feature of these three studies is that detailed description is lacking about the predominant type(s) of welding and the materials on which the welding is performed.

Welding is being carried out on a large scale in the industrialized world. Welders are exposed to noxious agents and materials such as oxides of iron, zinc, chromium, nickel, nitrogen dioxide, ozone, and asbestos. Many of these agents may be carcinogens, cocarcinogens, or promotors. In stainless steel welding, however, there is evidence[78-86] that hexavalent chromium compounds in the welding fumes may be of significant importance for the increased cancer risk associated with welding.

III. EXPERIMENTAL STUDIES IN ANIMALS

A. The Animal Model

Chromium compounds have been suspected of being carcinogenic in man since 1932, but which compounds are the most potent carcinogens has not been known. One of the purposes of animal studies, therefore, should be to differentiate between the carcinogenic potency of the different chromium compounds. So far, no human studies have demonstrated that exposure to trivalent chromium compounds is associated with increased risk of developing cancer.[45] Also, most animal studies have been carried out with hexavalent compounds.

The choice of the most relevant animal model for exposure with chromium compounds has been a great problem. Since the respiratory organs are the most common route of exposure in workers exposed to chromium compounds, inhalation exposure in animal models may be considered as the most adequate method. However, inhalation exposure in animals involves a number of difficulties, both in the design of the experiments and in the interpretation of the results.[91–93] Therefore, Kuschner and Laskin[94] developed a different method consisting of a stainless steel wire-mesh pellet, which served as carrier of different chromates diluted in cholesterol. The pellets were placed in the bronchus of the rats through a tracheostoma. Levy[95] tested a number of chromates in rats by means of this method. He found 7 epithelial bronchial tumors in 100 rats exposed to calcium chromates, and 3 tumors in 100 rats tested with zinc chromates. No tumors appeared in the control rats. Since this method implies a quite artificial manner of exposure, the interpretation of the results and its relevance ot humans is difficult.

Baetjer et al.[96] exposed mice and rats to mixed chromate dust at concentrations from 1 to 3 mg/m^3 for their whole lifetime. No lung tumors developed in these animals. The same mixed chromate dust was also administered intrapleurally, but yielded no tumors. The authors also administered basic zinc chromate and barium chromate intratracheally, but no tumors developed. In a later study, Steffee and Baetjer[97] exposed rabbits, guinea pigs, rats, and mice to zinc chromate, lead chromate, and mixed chromates. Repeated intratracheal dosing and inhalation exposure to 3 to 4 mg/m^3 was used. No malignant tumors in the respiratory tract were produced.

Nettesheim et al.[98] carried out an inhalation study exposing more than 1000 mice to calcium chromate 5 hr/day, 5 days a week for their whole lifetime. They found a significant increase in the number of adenocarcinomas and alveologenic adenomas in the exposed animals as compared with the control animals. No bronchiogenic tumors developed. This study is the only one reported in the literature demonstrating tumorigenic activity of a chromate with inhalation techniques as the method of exposure.

A number of studies have been carried out administering different chromium compounds intramuscularly, indicating the tumorogenic action of these compounds. Hueper and Payne[99] implanted calcium chromate, sintered calcium chromate, and sintered chromium trioxide in Bethesda Black strain rats. Sheep fat was used as vehicle, but it was indicated that the vehicle had no carcinogenic properties. More than 60% of the rats in each group developed tumors, mainly sarcomas and anaplastic giant cell sarcomas.

Roe and Carter[100] administered 20 once-weekly injections of calcium chromate intramuscularly in male rats, each rat receiving 19 mg calcium chromate. Spindle cell and pleomorphic cell sarcomas developed at the site of injection in 18 of the 24 rats. The mean development time was 323 days. The tumors did not metastasize. No tumors developed in the control rats which had received the vehicle only.

Schroeder et al.[101] administered 5 ppm trivalent chromium compounds to the diet of mice. The animals were fed the diet for their lifetime. No increase in the tumor incidence was observed as compared to the controls.

Langård et al.[102] exposed 3 groups of 24 rats 7 hr a day, 5 days a week for 4, 8, and 12 months, respectively, to 4 to 5 mg/m^3 zinc chromate dust. No bronchiogenic tumors developed in any of the groups.

B. Evaluation of the Results

Except for one study,[98] animal exposure to different chromate dusts has not confirmed the results found in human epidemiologic studies, where the association between exposure to chromates and lung cancer development has been quite convincing. No firm conclusion can be drawn from this apparent discrepancy. Two studies[8,102] are interesting in this respect. In one of the studies[8] 6 lung cancer cases had developed in a small human population of 24 workers who had been exposed to zinc chromate for more than 3 years sometime between 1948 and 1972. The workers had been exposed to atmospheric concentrations of zinc chromates from 0.25 to 1.5 mg/m^3. In all but one of the cases, a history of quite heavy smoking could be confirmed. In groups of Wistar rats exposed to about ten times as high concentrations of zinc chromates,[102] no bronchiogenic tumors developed. Although no final conclusions can be drawn from these studies, the combination of tobacco smoking and zinc chromate exposure in the human study, as opposed to zinc chromate exposure only in the animal study, may be an indication that the carcinogenic effect of zinc chromate is potentiated by the smoking, or vice versa. This could indicate the zinc chromate is acting as a cocarcinogen and not as a true carcinogen.

The studies with intramuscular administration of chromates give strong evidence that these chromium compounds cause cancer when administered by this route. However, this route of administration may be considered as being of limited interest for humans.

IV. CONCLUSION AND NEEDS FOR FURTHER RESEARCH

A. Importance of Water Solubility

The evidence for carcinogenic action of mono- and dichromates and chromium pigments is strong. The importance of the water solubility of chromates to carcinogenicity has been in focus for more than 30 years.[42-44] The carcinogenic action of the slightly soluble calcium chromate has been confirmed both in human epidemiologic studies and in different animal studies. On the other hand, the slightly soluble zinc chromate does cause bronchial cancer and possibly cancer in the gastrointestinal tract in exposed humans, while animal inhalation studies and installation of this compound in the bronchial tree in animals have not produced bronchiogenic cancers. Therefore, there are conflicting results on the importance of the solubility of these two compounds.

It has been suggested that insoluble lead chromate causes cancer in humans.[69,70] On the other hand, the highly soluble chromium trioxide may also cause cancer in humans.[24,25,74]

Consequently, the importance of the water solubility of hexavalent chromium compounds to carcinogenicity[42,43] may have been over emphasized as suggested elsewhere.[44] Other factors than the solubility of the inhaled compounds may be of much greater importance for the carcinogenic effects, such as the size distribution of the inhaled particles or mists, and factors which influence on the deposition and retention of the particles in the bronchial tree. In this respect, the influence of smoking on retention time in the bronchial tree may be of significant importance, both for the uptake and distribution of chromium compounds in the human body and on local concentration in the bronchial tree and the duration of exposure after interruption of exposure.

The total load of inhaled dust and other toxicants along with inhalation of chromates may be of significant importance to the efficacy of the clearance mechanisms in the bronchial tree. In the case of exposure to stainless steel welding smoke, where chromates are inhaled along with a number of other toxic agents, this factor may be of great importance to the carcinogenic action of the chromates present in the welding fumes.

B. Dose-Response Relationship

Only few epidemiologic studies have included atmospheric measurements.[34-39] Most of the measurements have been carried out at a time which is not necessarily relevant for the epidemiologic study and the available information on atmospheric concentrations deal with quite different compounds. Today, therefore, no conclusive extrapolation can be drawn from the high levels of exposure which have caused cancer in humans to low levels which may be considered to be "safe" for humans. However, except for the epidemiologic studies on welders, where the chromate concentration may be assumed to be lower than 50 μg (Cr)/m^3, no studies have been presented which for certainty document a carcinogenic hazard below this level. This lack of information should not lead to the conclusion that these levels are "safe" levels for the exposed workers. When considering this question, it is important to know that the epidemiologic methods which have been used are quite crude methods, particularly when the exposure characterization of the workers included in the studies is insufficient.

C. Research Needs

Epidemiologic studies which will enable us to estimate the cancer hazard at low chromate exposure levels are urgently needed. Legislating authorities want to set levels of exposure which are associated with minimal or no cancer hazard to the exposed worker. On the other hand, the authorities do not wish to set these levels at such a low level that the industry is unable to respond to the demands within acceptable costs. Therefore, epidemiologic studies on worker populations which have been or are exposed to low level chromate concentrations are needed.

Stainless steel welders may be adequate populations for this kind of study. The lower the exposure level, the more critical it becomes that an accurate occupational history for each individual is recorded and that measurements of the atmospheric concentrations of chromates are taken.

The importance of interaction between chromates and other exposure-factors for the development of cancer should also be studied in large groups. An accurate exposure characterization on different exposure factors in each individual may provide necessary information to enable the epidemiologist to determine whether chromates are true carcinogens or cocarcinogens.

Since the solubility of the different chromates has been considered of importance to carcinogenic potency, further studies should be carried out on workers who have been exposed to the soluble chromium trioxide.

Though the animal models have not been particularly successful in the case of exposure to chromates, inhalation exposure to chromic acid mists may help to determine whether this compound is carcinogenic. On the other hand, until most of the methodological difficulties with inhalation studies have been overcome, inhalation studies with chromates in particulate form should not be encouraged.

Needs for further research in the in vitro models are covered elsewhere in this volume.

REFERENCES

1. **Lehmann, K. B.**, Ist Grund zu einer besonderen Beunruhigung wegen des Auftretens von Lungenkrebs bei Chromatarbeitern vorhanden?, *Zentralbl. Gewerbehyg. Unfallverhuet.*, 19, 168, 1932.
2. **Newman, D.**, A case of adeno-carcinoma of the left inferior turbinated body, and perforation of the nasal septum, in the person of a worker in chrome pigments, *Glasgow Med. J.*, 33, 469, 1890.
3. **Alwens, W., Bauke, E. E., and Jonas, W.**, Auffallende Häufung von Bronchialkrebs bei Arbeitern der chemischen Industrie, *Arch. Gewerbepathol. Gewerbehyg.*, 7, 69, 1936.

4. **Pfeil, E.**, Lungentumoren als Berufserkrankung in Chromatbetrieben, *Dtsch. Med. Wochenschr.*, 61, 1197, 1935.
5. **Asang, E.**, Chronische Chromatschädigung mit Entwicklung eines Lungentumors, *Zentralbl. Arbeitsmed. Arbeitsschutz.*, 2, 181, 1952.
6. **Maclure, K. M. and MacMahon, B.**, An epidemiologic perspective of environmental carcinogenesis, *Epidemiol. Rev.*, 2, 19, 1980.
7. **Alderson, M. R., Rattan, N. S., and Bidstrup, L.**, Health of workmen in the chromate-producing industry in Britain, *Br. J. Ind. Med.*, 38, 117, 1981.
8. **Langård, S. and Vigander, T.**, Occurrence of lung cancer in workers producing chromium pigments, *Br. J. Ind. Med.*, in press.
9. **Schroeder, H. A., Balassa, J. J., and Tipton, I. H.**, Abnormal trace metals in man—chromium, *J. Chronic Dis.*, 15, 941, 1962.
10. **Kumpulainen, J. T., Wolf, W. R., Veillon, C., and Mertz, W.**, Determination of chromium in selected United States diets, *J. Agric. Food. Chem.*, 27, 490, 1979.
11. **Ayling, G. M. and Bloom, H.**, Heavy metals analyses to characterize and estimate distribution of heavy metals in dust fallout, *Atmos. Environ.*, 10, 61, 1976.
12. **Duce, R. A., Hoffman, G. L., and Zoller, W. H.**, Atmospheric trace metals at remote northern and southern hemisphere sites: pollution or natural?, *Science*, 187, 59, 1975.
13. **Crosmun, S. T. and Mueller, T. R.**, The determination of chromium (VI) in natural waters by differential pulse polarography, *Anal. Chim. Acta*, 75, 199, 1975.
14. **Fukai, R.**, Valency state of chromium in seawater, *Nature (London)*, 213, 901, 1967.
15. **Cary, E. E., Allaway, W. H., and Olson, O. E.**, Control of chromium concentrations in food plants. 1. Absorption and translocation of chromium by plants, *J. Agric. Food Chem.*, 25, 300, 1977.
16. **Gemmell, R. P.**, Revegetation of derelict land polluted by a chromate smelter, part 1: chemical factors causing substrate toxicity in chromate smelter waste, *Environ. Pollut.*, 5, 181, 1973.
17. **Fairbairn, J.**, Chromium pollution hits Japan, *New Sci.*, 67, 650, 1975.
18. **Axelsson, G. and Rylander, R.**, Environmental chromium dust and lung cancer mortality, *Environ. Res.*, 23, 469, 1980.
19. **Mertz, W.**, Chromium occurrence and function in biological systems, *Physiol. Rev.*, 49, 163, 1969.
20. **Langård, S. and Hensten-Pettersen, A.**, Chromium toxicology, in *Systemic Aspects of Biocompatibility*, Williams, D. F., Ed., CRC Press, Boca Raton, Fla., 1981, 143.
21. **Heath, J. C., Freeman, M. A. R., and Swanson, S. A. V.**, Carcinogenic properties of wear particles from prostheses made in cobalt-chromium alloy, *Lancet*, 1, 564, 1971.
22. **McKee, G. K.**, Carcinogenic properties of wear particles from prostheses made in cobalt-chromium alloy, (letter to the Editor) *Lancet*, 1, 750, 1971.
23. **Pedley, B., Meachim, G., and Williams, D. F.**, Tumor induction by implant materials, in *Fundamental Aspects of Biocompatibility*, Williams, D. F., Ed., CRC Press, Boca Raton, Fla., 1981, 175.
24. **Barbořík, M., Hanslian, L., Oral, L., Sehnalová, H., and Holusá, R.**, Carcinoma of the lungs in personnel working at electrolytic chromium plating, *Pracov. Lék.*, 10, 413, 1958; English abstract in *Excerpta Med. Sect. 16 Cancer*, 7, 1395, 1959.
25. **Royle, H.**, Toxicity of chromic acid in the chromium plating industry (1), *Environ. Res.*, 10, 39, 1975.
26. **Michel-Briand, C. and Simonin, M.**, Cancers broncho-pulmonaires survenus chez deux salariés occupés à un poste de travail dans le même atelier de chromage electrolytique, *Arch. Mal. Prof. Med. Trav. Secur. Soc.*, 38, 1001, 1977.
27. **Petrilli, F. L. and de Flora, S.**, Toxicity and mutagenicity of hexavalent chromium on *Salmonella typhimurium*, *Appl. Environ. Microbiol.*, 33, 805, 1977.
28. **Ormos, G. and Mányai, S.**, Chromate uptake by human red blood cells: comparison for different divalent anions, *Acta Biochim. Biophys. Acad. Sci. Hung.*, 9, 197, 1974.
29. **Bazerbashi, M. B., Reeve, J., and Chanarin, I.**, Studies in chronic lymphocytic leukemia. The kinetics of ^{51}Cr-labelled lymphocytes, *Scand. J. Haematol.*, 20, 37, 1978.
30. **Davey, M. G. and Lander, H.**, The labelling of human platelets with radiochromate, *Aust. J. Exp. Biol. Med. Sci.*, 41, 581, 1963.
31. **McCaul, T. F., Poston, R. N., and Bird, R. G.**, *Entamoeba histolytica* and *Entamoeba invadens*: chromium release from labeled human liver cells in culture, *Exp. Parasitol.*, 43, 342, 1977.
32. **Dybdahl, J. H., Daae, L. N. W., Larsen, S., Ekeli, H., Frislid, K., Wiik, I., and Aanstad, L.**, Acetylsalicylic acid-induced gastrointestinal bleeding determined by a ^{51}Cr method on a day-to-day basis, *Scand. J. Gastroenterol.*, 15, 887, 1980.
33. **Järvholm, B., Lillienberg, L., Sällsten, G., and Thiringer, G.**, Method of the cohort study—a practical example, *Eur. J. Respir. Dis.*, 61(Suppl. 107), 99, 1980.

34. **Machle, W. and Gregorius, F.**, Cancer of the respiratory system in the United States chromate-producing industry, *Public Health Rep.*, 63, 1114, 1948.
35. **Mancuso, T. F. and Hueper, W. C.**, Occupational cancer and other health hazards in a chromate plant: a medical appraisal. 1. Lung cancers in chromate workers, *Ind. Med. Surg.*, 20, 358, 1951.
36. **Mancuso, T. F.**, Occupational cancer and other health hazards in a chromate plant: a medical appraisal. II. Clinical and toxicologic aspects, *Ind. Med. Surg.*, 20, 393, 1951.
37. **Langård, S. and Norseth, T.**, A cohort study of bronchial carcinomas in workers producing chromate pigments, *Br. J. Ind. Med.*, 32, 62, 1975.
38. **Langård, S., Andersen, A., and Gylseth, B.**, Incidence of cancer among ferrochromium and ferrosilicon workers, *Br. J. Ind. Med.*, 37, 114, 1980.
39. **Axelsson, G., Rylander, R., and Schmidt, A.**, Mortality and incidence of tumours amoung ferrochromium workers, *Br. J. Ind. Med.*, 37, 121, 1980.
40. **Baetjer, A. M.**, Pulmonary carcinoma in chromate workers, 1. a review of the literature and report of cases, *AMA Arch. Ind. Hyg. Occup. Med.*, 2, 487, 1950.
41. **Gafafer, W. M., Ed.**, Health of Workers in Chromate Producing Industry. A Study, Public Health Service Publication No. 192, U.S. Government Printing Office, Washington, D.C., 1952.
42. NIOSH, Criteria for a Recommended Standard: Occupational Exposure to Chromic Acid, Department of Health, Education and Welfare (NIOSH) Publ. No. 73-11021, National Institute for Occupational Safety and Health, Cincinnati, Ohio, 1973.
43. NIOSH, Criteria for a Recommended Standard: Occupational Exposure to Chromium VI, Department of Health, Education and Welfare (NIOSH) Publ. No. 76-129, National Institute for Occupational Safety and Health, Cincinnati, Ohio, 1975.
44. Nordiska Expertgruppen för Gränsvärdesdokumentation, 8. Krom, *Arbete och Hälsa*, No. 33, 1979.
45. **Langård, S.**, Chromium, in *Metals in the Environment*, Waldron, H. A., Ed., Academic Press, London, 1980, chap. 4.
46. **Enterline, P. E.**, Respiratory cancer among chromate workers, *J. Occup. Med.*, 16, 523, 1974.
47. **Langård, S. and Norseth, T.**, Chromium, in *Handbook on the Toxicology of Metals*, Friberg, L., Ed., Elsevier/North Holland Biomedical Press, Amsterdam, 1979, chap. 22.
48. NAS, Committee on Biological Effects of Atmospheric Pollutants, Chromium, National Academy of Sciences, Washington, D.C., 1974.
49. **Norseth, T.**, The carcinogenicity of chromium, *Environ. Health Perspect.*, 40, 121, 1981.
50. **Hayes, R. B.**, Cancer and occupational exposure to chromium chemicals, *Rev. Cancer Epidemiol.*, 1, 293, 1980.
51. International Agency for Research on Cancer, Some metals and metallic compounds, *IARC Monogr. Eval. Carcinog. Risk Chem. Man*, 23, 205, 1980.
52. **Léonard, A. and Lauwerys, R. R.**, Carcinogenicity and mutagenicity of chromium, *Mutat. Res.*, 76, 227, 1980.
53. **Carozzi, L.**, Cancer professionnel et organisation internationale du travail, *Acta Unio. Int. Cancrum*, 2, 3, 1937.
54. **Alwens, W. and Jonas, W.**, Der Chromat-Lungenkrebs, *Acta Unio. Int. Cancrum*, 3, 103, 1938.
55. **Bloomfield, J. J. and Blum, W.**, Health hazards in chromium plating, *Public Health Rep.*, 43, 2330, 1928.
56. **Hueper, W. C.**, Occupational and environmental cancers of the respiratory system, *Recent Results Cancer Res.*, 3, 56, 1966.
57. **Bidstrup, P. L.**, Carcinoma of the lung in chromate workers, *Br. J. Ind. Med.*, 8, 302, 1951.
58. **Bidstrup, P. L. and Case, R. A. M.**, Carcinoma of the lung in workmen in the bichromates-producing industry in Great Britain, *Br. J. Ind. Med.*, 13, 260, 1956.
59. **Baetjer, A. M.**, Pulmonary carcinoma in chromate workers. II. Incidence on basis of hospital records, *AMA Arch. Ind. Hyg. Occup. Med.*, 2, 505, 1950.
60. **Taylor, F. H.**, The relationship of mortality and duration of employment as reflected by a cohort of chromate workers, *Am. J. Public Health*, 56, 218, 1966.
61. **Ohsaki, Y., Abe, S., Kimura, K., Tsuneta, Y., Mikami, H., and Murao, M.**, Lung cancer in Japanese chromate workers, *Thorax*, 33, 372, 1978.
62. **Hayes, R. B., Lilienfeld, A. M., and Snell, L. M.**, Mortality in chromium chemical production workers: a prospective study, *Int. J. Epidemiol.*, 8, 365, 1979.
63. **Pokrovskaya, L. V. and Shabynina, N. K.**, Carcinogenous hazards in the production of chromium ferroalloys, *Gig. Tr. Prof. Zabol.*, 10, 23, 1973 (In Russian, English abstract).
64. **Gross, E. and Kölsch, F.**, Über den Lungenkrebs in der Chromfarbenindustrie, *Arch. Gewerbepathol. Gewerbehyg.*, 12, 164, 1943.
65. **Stokinger, H. E.**, The metals: 9. chromium, in *Patty's Industrial Hygiene and Toxicology*, Vol. 2A, 3rd rev. ed., Clayton, G. D. and Clayton, F. E., Eds., John Wiley & Sons, New York, 1981, 1589.

66. **Langård, S. and Andersen, A.**, Betydningen av Bruk av Latenstid, Kritisk Alder og Valg av Referansepopulasjon ved Beregning av Cancerinsidens i Kohort-studier, presented at 29. Nordiske Yrkeshygieniske Møte i Norge, November 1980, Rep. HD 842, Yrkeshygienisk Institutt, Oslo, 1980.
67. **Langård, S. and Norseth, T.**, Cancer in the gastrointestinal tract in chromate pigment workers, *Arh. Hig. Rada Toksikol.*, 30(Suppl.), 301, 1979.
68. Equitable Environmental Health Inc., An Epidemiological Study of Lead Chromate Plants: Final Report, prepared for The Dry Color Manufacturers Association, 1976.
69. **Davies, J. M.**, Lung-cancer mortality of workers making chrome pigments, *Lancet*, 1, 384, 1978.
70. **Davies, J. M.**, Lung cancer mortality of workers in chromate pigment manufacture: an epidemiological survey, *J. Oil Colour Chem. Assoc.*, 62, 157, 1979.
71. **Frentzel-Beyme, R. and Claude, J.**, Lung cancer mortality in workers employed in chromate pigment factories. A multicentric Central European epidemiologic study, *Am. J. Epidemiol.*, 112, 423, 1980.
72. **Dalager, N. A., Mason, T. J., Fraumeni, J. F., Hoover, R., and Payne, W. W.**, Cancer mortality among workers exposed to zinc chromate paints, *J. Occup. Med.*, 22, 25, 1980.
73. **Chiazze, L., Ference, L. D., and Wolf, P.**, Mortality among automobile assembly workers. I. Spray painters, *J. Occup. Med.*, 22, 520, 1980.
74. **Waterhouse, J. A. H.**, Cancer among chromium platers, *Br. J. Cancer*, 32, 262, 1975.
75. **Okubo, T. and Tsuchiya, K.**, Epidemiological study of chromium platers in Japan, *Biol. Trace Elem. Res.*, 1, 35, 1979.
76. **Sehnalová, H. and Barbořík, M.**, K otázce ukládání sloučenin chrómu u lidí, (On the problem of deposition of chromium compounds in man), *Prac. Lék.*, 17, 399, 1965.
77. **Petrilli, F. L. and de Flora, S.**, Metabolic deactivation of hexavalent chromium mutagenicity, *Mutat. Res.*, 54, 139, 1978.
78. **Fregert, S. and Övrum, P.**, Chromate in welding fumes with special reference to contact dermatitis, *Acta Derm. Venereol.*, 43, 119, 1963.
79. **Naranjit, D., Thomassen, Y., and van Loon, J. C.**, Development of a procedure for studies of the chromium(III) and chromium(VI) contents of welding fumes, *Anal. Chim. Acta*, 110, 307, 1979.
80. **Thomsen, E. and Stern, R. M.**, A simple analytical technique for the determination of hexavalent chromium in welding fumes and other complex matrices, *Scand. J. Work Environ. Health*, 5, 386, 1979.
81. **White, L. R., Jakobsen, K., and Østgaard, K.**, Comparative toxicity studies of chromium-rich welding fumes and chromium on an established human cell line, *Environ. Res.*, 20, 366, 1979.
82. **Hedenstedt, A., Jenssen, D., Lidesten, B.-M., Ramel, C., Rannug, U., and Stern, R. M.**, Mutagenicity of fume particles from stainless steel welding, *Scand. J. Work Environ. Health*, 3, 203, 1977.
83. **Maxild, J., Andersen, M., Kiel, P., and Stern, R. M.**, Mutagenicity of fume particles from metal arc welding on stainless steel in the *Salmonella*/microsome test, *Mutat. Res.*, 56, 235, 1978.
84. **White, L. R., Hunt, J., Tetley, T. D., and Richards, R. J.**, Biochemical and cellular effects of welding fume particles in the rat lung, *Ann. Occup. Hyg.*, 24, 93, 1981.
85. **Gottlieb, M. S.**, Lung cancer and the petroleum industry in Louisiana, *J. Occup. Med.*, 22, 384, 1980.
86. **Sjögren, B.**, A retrospective cohort study of mortality among stainless steel welders, *Scand. J. Work Environ. Health*, 6, 197, 1980.
87. **Breslow, L., Hoaglin, L., Rasmussen, G., and Abrams, H. K.**, Occupations and cigarette smoking as factors in lung cancer, *Am. J. Public Health*, 44, 171, 1954.
88. **Milham, S.**, Cancer mortality patterns associated with exposure to metals, *Ann. N.Y. Acad. Sci.*, 271, 243, 1976.
89. **Menck, H. R. and Henderson, B. E.**, Occupational differences in rates of lung cancer, *J. Occup. Med.*, 18, 797, 1976.
90. **Beaumont, J. J. and Weiss, N. S.**, Mortality of welders, shipfitters, and other metal trades workers in boilermakers local No. 104, AFL-CIO, *Am. J. Epidemiol.*, 112, 775, 1980.
91. **Phalen, R. F.**, Inhalation exposure of animals, *Environ. Health Perspect.*, 16, 17, 1976.
92. **Campbell, K. I.**, Inhalation toxicology, *Clin. Toxicol.*, 9, 849, 1976.
93. **Langård, S. and Nordhagen, A.-L.**, Small animal inhalation chambers and the significance of dust ingestion from the contaminated coat when exposing rats to zinc chromate, *Acta Pharmacol. Toxicol.*, 46, 43, 1980.
94. **Kuschner, M. and Laskin, S.**, Experimental models in environmental carcinogenesis, *Am. J. Pathol.*, 64, 183, 1971.

95. **Levy, L. S.**, Effects of Various Chromium-containing Materials on Rat Lung Epithelium, Ph.D. thesis, Institute of Cancer Research, University of London, London, 1975.
96. **Baetjer, A. M., Lowney, J. F., Steffee, H., and Budacz, V.**, Effect of chromium on incidence of lung tumors in mice and rats, *A.M.A. Arch. Ind. Health,* 20, 124, 1959.
97. **Steffee, C. H. and Baetjer, A. M.**, Histopathologic effects of chromate chemicals. Report of studies in rabbits, guinea pigs, rats, and mice, *Arch. Environ. Health,* 11, 66, 1965.
98. **Nettesheim, P., Hanna, M. G., Doherty, D. G., Newell, R. F., and Hellman, A.**, Effect of calcium chromate dust, influenza virus, and 100 R whole-body X radiation on lung tumor incidence in mice, *J. Natl. Cancer Inst.,* 47, 1129, 1971.
99. **Hueper, W. C. and Payne, W. W.**, Experimental cancers in rats produced by chromium compounds and their significance to industry and public health, *Am. Ind. Hyg. Assoc. J.,* 20, 274, 1959.
100. **Roe, F. J. C. and Carter, R. L.**, Chromium carcinogenesis: calcium chromate as a potent carcinogen for the subcutaneous tissues of the rat, *Br. J. Cancer,* 23, 172, 1969.
101. **Schroeder, H. A., Balassa, J. J., and Vinton, W. H.**, Chromium, lead, cadmium, nickel and titanium in mice: effect on mortality, tumors and tissue levels, *J. Nutr.,* 83, 239, 1964.
102. **Langård, S., Lexow, P. B., and Nordhagen, A.-L.**, The carcinogenic potency of zinc chromate in Wistar rats. A long-term inhalation study, presented at 30. Nordiska Yrkeshygieniska Mötet, Åbo, Finland, Oktober 1981, Institute of Occupational Health, Helinski, Finland, 1981.

Chapter 3

EFFECTS OF CHROMIUM COMPOUNDS ON THE RESPIRATORY SYSTEM

P. L. Bidstrup

TABLE OF CONTENTS

I.	Introduction	32
II.	Ulceration and Perforation of the Nasal Septum	32
III.	Bronchospasm	34
IV.	Asthma	34
V.	Lung Cancer	35
	A. Epidemiology	35
	1. Early Diagnosis of Lung Cancer	36
VI.	Mutagenicity	41
VII.	Induction of Tumors in Experimental Animals	45
References		49

I. INTRODUCTION

Only hexavalent compounds of chromium affect the respiratory system. Korallus et al.[1] have reported the results of investigations on 106 men exposed only to trivalent (Cr.3) compounds—64 for more than 10 years. They found no evidence of an increased incidence of lung cancer, airways obstruction, nasal septal ulceration, or dermatitis. In the group, there were six who had radiological evidence of pneumoconiosis but in five there was a previous history of exposure to silica in mining, sand-blasting, or tunnel construction and the sixth case was assessed, by an independent radiologist, as "possibly suspicious."

Exposure to dust or mist containing hexavalent chromium may cause:

1. Ulceration, leading to perforation, of the nasal septum
2. Chemical irritation of the bronchial mucous membrane resulting in bronchospasm
3. Asthma due to sensitization
4. Carcinoma of lung in those engaged in the manufacture of hexavalent chromium compounds from the ore, chromite (which is trivalent), and possibly in the manufacture of some chrome pigments

II. ULCERATION AND PERFORATION OF THE NASAL SEPTUM

There are many causes, other than hexavalent chromium of ulceration and perforation of the nasal septum. The most common is trauma secondary to nose-picking, surgical procedures such as submucous resection, and hematoma following a blow on the nose. Other causes are exposure to anhydrous sodium carbonate (soda ash); arsenic and its compounds; organic compounds of mercury, particularly mercury fulminate; alkaline dusts such as soap powders; hydrofluoric acid and fluorides; capsaicin, the pungent active principle of capsicum (chillies); vanadium; dimethyl sulfate; cocaine and other drugs taken as snuff; copper salts (rarely); lime (rarely); lupus or syphilitic infections; neoplasms.

Among many descriptions of nasal ulceration due to hexavalent compounds, the experience of doctors employed in the chromates-producing industry in Great Britain is summarized in an annotation to *British Medical Journal*.[2] Their summated experience of chrome ulceration covered 100 years.

Soluble hexavalent chromium compounds are irritant and corrosive. Inhalation of dust or mist containing hexavalent chromium, or direct application by the finger in those whose personal hygiene is of low standard, will result in irritation of the nasal mucous membrane as a whole with localized effects over the lower, anterior part of the septum—Kiesselbach's or Little's area—which is relatively avascular and closely adherent to the underlying cartilage. It is also the site at which incoming streams of air impinge on the septum.

There are four well-marked stages in the progression from nasal irritation to ulceration and perforation of the septum. Redness and irritation of the mucosa covering the septum is associated with rhinorrhea followed by blanching over the lower, anterior part. Tough adherent crusts form causing discomfort and encourage the tendency to remove them quickly by nose-picking. This may cause abrasion and further contamination with hexavalent chromium. After a period, which may vary from days to weeks, ulceration occurs and the crusts which continue to form include necrotic debris from the cartilage of the septum.

This stage is followed by erosion of the septal cartilage resulting in a perforation, the periphery of which is still the site of active ulceration (Figure 1). The perforation

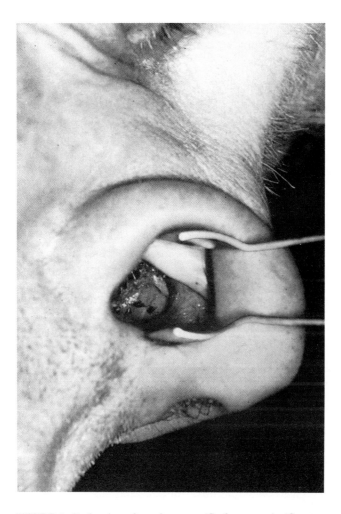

FIGURE 1. Perforation of nasal septum (Cr 6 exposure). (Courtesy of Prof. I. McCallum.)

may increase in size, extending upwards and posteriorly, but neither the bony part of the septum nor the anterior and lower parts become involved. Thus the shape of the nose is not altered.

In the stage of active ulceration slight nose-bleed and pain at the bridge of the nose on contact with cold air are common symptoms. The edges of the perforation heal by formation of avascular scar tissue, usually within 3 months, but healing may be delayed by the trauma of nose-picking. In many cases, once active ulceration has healed, there are no symptoms but a tendency to "wet nose" and crusting may persist and when the perforation is of small diameter, a whistling on inspiration may cause embarrassment.

It is difficult to prevent nasal septal ulceration in those at risk of inhaling hexavalent compounds at work. Experience in the chromates-producing industry in Great Britain, in which soluble hexavalent chromium compounds are made by converting trivalent chromium in the ore chromite to the hexavalent form, has confirmed that a high standard of personal hygiene is a major factor in prevention and that control of sources of exposure to dust and mist by appropriate design of process will reduce the risk considerably. Routine static and personal air monitoring are recommended.

All employees must be made aware of the potential risk. New entrants must be instructed in matters of hygiene and all must report for treatment at the earliest sign of nasal irritation. Treatment using an ointment containing 10% sodium calcium edetate has proved effective in promoting prompt relief of nasal irritation and, even when this proceeds to ulceration, it is unusual for perforation of the septum to occur. It is most important to persuade the patient not to remove crusts by digital manipulation since this delays healing and enhances the risk of perforation.

Chrome ulceration of the skin was described by Cumin[3] in 1827 and in 1863, Bécourt and Chevallier[4] reported ulceration of both skin and nasal mucous membrane. It is not possible to estimate the number of persons who have developed nasal septal ulceration or perforation due to exposure to hexavalent chromium compounds since these are manufactured and used in a variety of industries in many countries.

There is no evidence that the risk of developing nasal cancer is greater than expected in those who have suffered nasal ulceration or perforation due to hexavalent chromium compounds. The number of cases reported in the world literature is remarkably small. In a follow-up study of 2715 men employed in the chromates-producing industry in Great Britain for at least one year from 1948 to 1977, two deaths from nasal cancer were recorded.[5] Only one other is known to have occurred in some 6000 employees in this industry for whom records are available from 1955.

III. BRONCHOSPASM

Since 1955, employees in the chromates-producing industry in Great Britain have been under close medical supervision which includes preemployment medical examination with chest X-ray and examination by the works doctor after absences from work due to illness or injury for a period of 5 days or more. It became clear in the late 1950s that a number of men were absent because "bronchitis" had been diagnosed, but, as records accumulated, it was noted that such absences were for short duration—at most 4 to 5 days. It is more likely than not that the cause of the so-called "bronchitis" was chemical irritation of the lower respiratory tract, causing bronchospasm.

Airborne dust or mist containing hexavalent chromium is irritating to skin and mucous membranes and as sources of escape of dust or mist into the general working atmosphere were identified and controlled, short-term absences due to bronchial irritation became rare.

IV. ASTHMA

The potential of hexavalent chromium compounds to cause skin sensitization is well-known. Although asthma may occur, it is rare. Usually symptoms begin soon after starting work with hexavalent chromium compounds, quickly subside but recur on resuming work. It is assumed, although not proven, that the employee has become sensitized to chromates and he is advised to seek other employment.

It has been suggested that the incidence of chronic bronchitis may be increased in those at risk of inhaling dust or mist containing hexavalent chromium. Alderson et al.[5], reported on follow-up of 2715 men employed for one year or more from 1948 to 1977 in the chromates-producing industry in Great Britain. The average number of person-years was 16.3 (range 19.1 to 14.5). Two factories were closed in the late 1960s and the follow-up period for ex-employees is longer whereas, for the factory which is still in production, individuals have been included who entered the industry in the late 1960s and early 1970s. The follow-up revealed no evidence of an increased mortality from chronic bronchitis in the group as a whole but employees from the factory in Scotland did, in fact, have a higher than expected incidence (observed deaths 31 : expected

17.4). Deaths from other causes were higher, also, in this group. For the other two factories, one of which is still in production, observed deaths from chronic bronchitis were less than expected.

V. LUNG CANCER

A. Epidemiology

An increased risk of lung cancer among men employed in the manufacture of bichromates from the ore, chromite, was established for the industry in Germany in the 1930s, in the U.S. in the 1940s, in Great Britain in the 1950s, and in Japan in the 1970s. Relevant reports are Machle and Gregorius,[6] Baetjer,[7] Bidstrup and Case,[8] and Satoh et al.[9]

In 1950, Långard and Norseth[10] reported results of a cohort study of workers manufacturing chrome pigments in Norway. They found three cases of bronchogenic carcinoma (expected 0.079) but the numbers in the cohort were small and interpretation of the results difficult. The results suggest, but do not confirm, that there may be an increased risk of lung cancer in those engaged in the manufacture of chrome pigments. Exposure was to the raw materials used in manufacture and to the chromates of lead and zinc. Davies[11] reported preliminary results of a mortality study undertaken among men employed on manufacture of chrome pigments in three English factories. In two, both lead and zinc chromate were manufactured; only lead chromate was produced in the third. There was a significant excess of deaths from lung cancer among men working on both lead and zinc chromates, but further analysis reported by Davies,[12] suggested that, under conditions prevailing since 1946 at the factory where only lead chromate was produced, there was no increased risk of lung cancer.

Further epidemiological studies are needed to establish whether or not there is an increased risk in other industries where chrome-containing compounds are used—such as chromium-plating, spray-painting, and tanning of leather. Chiazze[13] has reviewed recent literature on the epidemiology of respiratory cancer and other health effects among workers exposed to chromates. He confirms the established risk of lung cancer in men engaged in chromates-manufacture but concludes, in relation to the user industries—"Studies on workers exposed to chromium compounds outside the chromate-producing industry, especially plating and spray-painting, generally do not point to an excess respiratory cancer risk. There may, however, be an excess risk for those engaged in chromate pigment production."

Where lung cancer has been recognized as a potential hazard of employment in the chromates-producing industry, every effort has been made to eliminate or reduce the risk. Recent follow-up studies have indicated reduction and possibly elimination of the risk for a chromates-producing plant operating in Baltimore, Md. since 1845,[14,15] in the industry in West Germany,[16] and in Great Britain.[5]

The investigations relating to the lung cancer problem in the British chromates-producing industry illustrate the measures which have been taken to identify the problem and attempt to overcome it. Following the report of Machle and Gregorius[6] relating to the lung cancer risk in the industry in the U.S. and referring to the similar risk already established in Germany, Bidstrup[17] began an investigation into the risk, or possible risk, in the industry in Great Britain. No past records were available—the survey was made, therefore, by clinical examination of those at risk, including detailed occupational history and chest X-ray from which has developed a prospective study which is continuing. The survey revealed one case of lung cancer; the expected was 0.41. By 1956, Bidstrup and Case[8] had found 12 cases; the expected was 3.3.

Analysis of data (unpublished) continued. It was not possible at that time to trace those who had left the industry to seek other work and follow-up was limited to those currently employed and those who had retired. In recent years, all data relating to men who have been employed at any time in the industry have been put onto computer in the Division of Epidemiology, Institute of Cancer Research, Sutton, Surrey. Analysis of data relating to 2715 men who worked in the industry for one year or more between 1948 and 1977 has been completed recently.[5] Only 298 of the total population complying with these terms have been lost to follow-up and the average person-years in the study was 16.3.

From the early 1950s many changes were made in working methods and in control of escape of chrome-containing dusts or mists into the working environment. A major change in process was introduced in 1961 at the works still in production—lime, previously added to the ore in the first stage of the process, was eliminated, thus removing the risk of exposure to slowly-soluble hexavalent chromium (calcium chromate) in the roast from the kilns.

Two factories ceased production in the 1960s, Alderson et al.[5] therefore made separate analysis of data relating to those within the total group who were, or had been, employed at the remaining production site. It was possible to distinguish three groups:

1. Men who had worked only before major plant modification
2. Men who had worked throughout the period of that modification
3. Men who began only after modification had been completed

For those entering the industry after plant modification, including change of process, the completed results of follow-up show a significant decrease in relative risk and suggest that the risk may have been eliminated. The follow-up is continuing because the time-interval for those employed only since modification of work methods, improved industrial hygiene, and change of process is relatively short (1961 to 1977) and it is possible that these changes have lengthened the latent period and have not eliminated the health risk. For the whole group, 116 deaths from lung cancer were observed and 48 expected (ratio observed/expected 2.4). Of the 116 cases observed, 60 worked in the industry for more than 25 years.

The study referred to[5] has confirmed the findings of previous studies with regard to the bichromates-producing industry in Great Britain, and in all other countries for which data are available, that there is no increase in relative risk of cancer at sites other than the lung, nor is there evidence of increased mortality from other causes.

1. Early Diagnosis of Lung Cancer

Since 1955, a preemployment medical examination (including chest X-ray) and medical examination after a sickness absence of 5 or more days has been standard procedure in the chromates-producing industry in Great Britain. Annual chest X-rays on 14 × 17 film was arranged for all employees and pensioners. Arrangements were made for those who had left the industry to seek other employment also to be X-rayed annually, but response from that group has been poor. The Works Medical Officer may arrange a chest X-ray, or other investigation, at any time.

In the early 1960s, it became apparent that X-rays at intervals of 12 months failed to reveal lung cancer at an early stage.[18] The decision was taken to reduce the interval from 12 to 8 months. For those who attend regularly for chest X-rays, lung cancer can be diagnosed and treated at a very early stage and for those treated promptly, usually by operation, a 5-year survival rate is 33% and for the whole group (which includes many in whom small shadows on chest X-rays were "observed" and operation delayed) 9%.[19]

Small shadows on X-rays, usually at the periphery, are seen before sputum cytology or bronchoscopy give positive results and in most cases the patient is symptom-free or unaware of cough or other symptoms until questioned directly about them. In many cases, local resection or lobectomy has resulted in survival beyond 5 years and has been compatible with normal or near-normal working capacity. Excretion of chromium in urine or chromium content of lung tissue confirm absorption of chromium-containing materials but are similar in those who died from lung cancer or other unrelated causes.[18] Tumors of all cell types were found, the majority being squamous-cell carcinoma.

In the early days of the investigation, particularly for those who died before 1960 (having left the industry to go to other jobs), information came late and often by death certificate only. In many who developed lung cancer, operation was not contemplated and post-mortem examination not undertaken as a routine. Thus cell-type of tumor and survival time after diagnosis is not known for all cases of lung cancer. It was possible to establish, however, that more than one third had worked on a process stage where exposure to slowly-soluble hexavalent chromium would have been in the roast, resulting from passing ore (mixed with lime and soda ash) through kilns to convert the insoluble chromium in chromite (trivalent) to soluble hexavalent chromium. Some calcium chromate, which is hexavalent and slowly soluble, would be formed and this compound has been shown to be carcinogenic to animals subjected to subcutaneous injection or intrabronchial pellet implantation.[30,31,34]

The importance of early diagnosis and prompt treatment, usually by operation, is illustrated in the following brief case records.

Case 1—A man who had worked from 1939 to 1942 and returned to the industry in 1955, was found (at routine X-ray) to have a large mass in the right lung. He admitted to no symptoms but agreed to an operation; pneumonectomy revealed a squamous-cell carcinoma. In 1981, he was 77 years old.

Case 2—Another, now aged 68, had left pneumonectomy for squamous-cell carcinoma in 1954. He resumed work until age 65; there has been no recurrence.

Case 3—In 1960, a maintenance fitter who had been employed on process for a short time in 1952, returning as a fitter in 1957, was found to have a shadow at the left apex (Figures 2A and 2B). He was referred to further investigation and treated for one year for tuberculosis. Another X-ray in 1961 at the annual works X-ray survey confirmed extension of the shadowing and left penumonectomy was necessary. The tumor was a squamous-cell carcinoma. He resumed work as joiner in a nonrisk area, working only on clean materials to age 65. A second primary tumor developed in the right lung, causing death 14 years after operation and 1 year after retirement.

Case 4—A man in his early 60s was taken on as a joiner's mate and worked with the employee referred to above. He would not have had exposure to chromium compounds. In 1964, 3 years later, a small shadow was found in the right mid-zone at routine works X-ray examination (Figures 3A and 3B). Local resection established that this was caused by a small, undifferentiated carcinoma. He died 13 years later at age 77 of a massive myocardial infarction. Post-mortem examination failed to reveal any evidence of recurrence or secondary spread.

Case 5—In 1968, a small shadow appeared in the right upper lobe of an employee who had worked in the industry for 16 years (Figures 4A and 4B). Right upper lobectomy confirmed squamous-cell carcinoma; survival time was 10 years.

Case 6—An employee with 33 years service died at age 71 from coronary thrombosis 7 years after right upper lobectomy had revealed an undifferentiated carcinoma. No recurrence or secondary spread was found at post-mortem examination.

These are among many examples where prompt operation without confirmation of diagnosis by means other than X-ray has proved successful and has enabled a man to

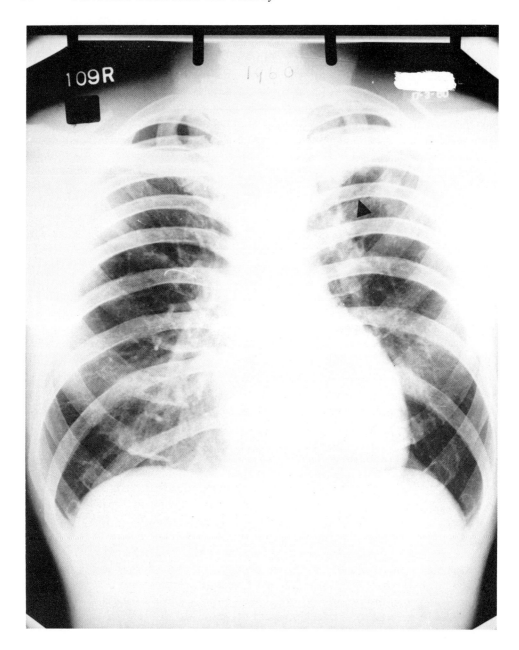

A

FIGURE 2. (A) Left apical shadow; squamous-cell carcinoma; (B) macroradiograph of 2A. Survival 14 years (Case 3).

resume work. By contrast, if small shadows on X-rays are overlooked or time is lost waiting for confirmation of diagnosis by bronchoscopy, sputum cytology, or other means, chance of cure is greatly reduced.

Case 7—Figures 5A and 5B illustrate the above in a man employed for 34 years. Although the small shadow at the mid-point of the right diaphragm was present in 1957,

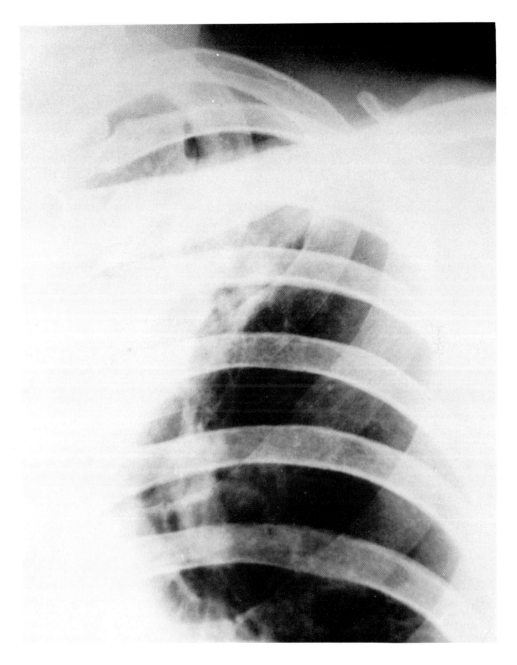

FIGURE 2B

no action was taken until 1958 (Figure 5B). Survival time was less than 1 year.

Case 8—Figures 6A and 6B show a mid-zone shadow which proved to be squamous-cell carcinoma. The man survived right upper lobectomy for 5 years.

When data were being collected for the original survey from 1948 to 1949,[17] the smoking-lung cancer association had not been reported. Thus, smoking habits were not recorded at first interview, but the information was sought subsequently by questionnaire. There is no evidence that the smoking habits of those taking part in the survey

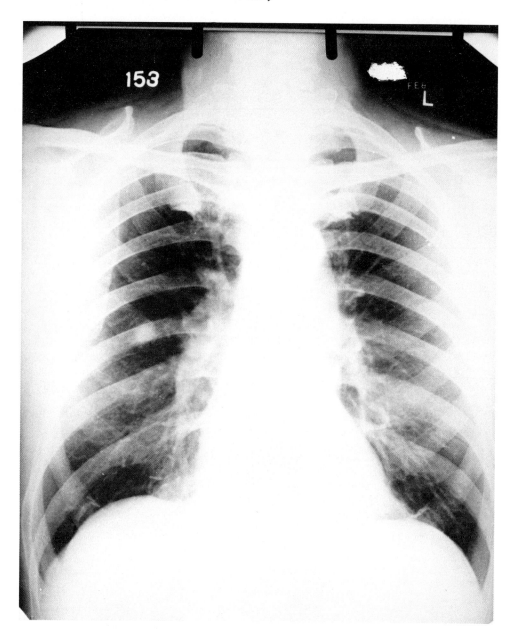

A

FIGURE 3. (A) Right mid-zone shadow, undifferentiated carcinoma; (B) macroradiograph of 3A. Survival 13 years. Death from myocardial infarction. No recurrence (Case 4).

differed from those of the male population in general. Reliable records on smoking habits exist for all employees from 1961 and will be collated with other data relating to men still employed on the remaining production site in the report which will follow that of Alderson et al.[5] Results to date suggest that the lung cancer risk has been eliminated for those employed from 1961, but the possibility that the latent interval has been extended must be excluded by continuing to analyze data relating to those who have been employed in the industry.

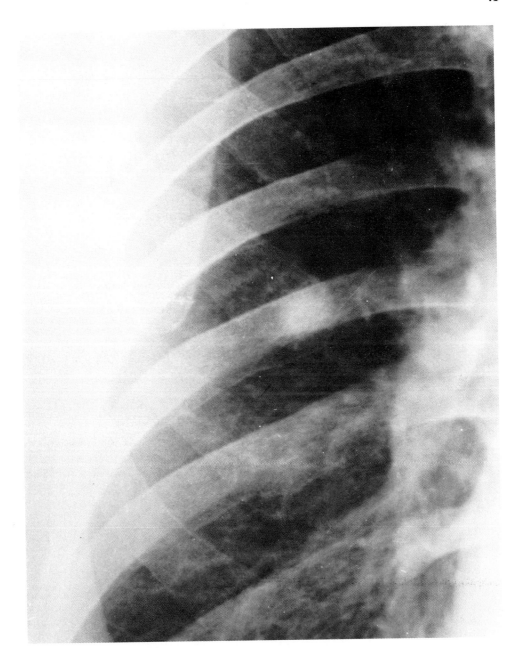

FIGURE 3B

VI. MUTAGENICITY

All hexavalent compounds which have been tested are mutagenic but not necessarily carcinogenic (see below). The activity depends upon solubility and because they are toxic, the range of activity is narrow. Trivalent compounds have no mutagenic potential; where positive results have been obtained, contamination of the test substance with hexavalent chromium has been established in each case.

De Flora,[20] has reported an investigation of 106 organic and inorganic compounds in the *Salmonella*/microsome test, including 11 hexavalent and 7 trivalent compounds.

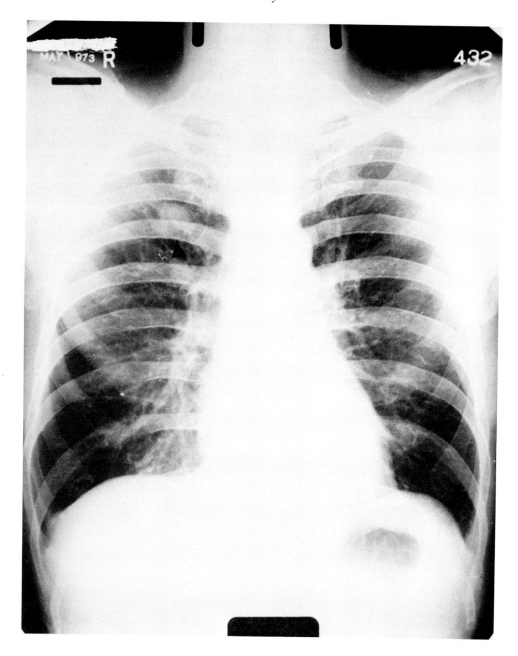

A

FIGURE 4. (A) Right upper zone shadow, poorly differentiated squamous carcinoma; (B) macroradiograph of 4A. Survival 7 years. Death from coronary thrombosis. No recurrence (Case 5).

In an earlier paper, Petrilli and De Flora[21] recorded that their results suggest that hexavalent chromium induces both frameshift errors and base-pair substitutions in *Salmonella* DNA.

Reducing chemicals, such as ascorbic acid and sodium sulfite or metabolites (NADH, NADPH, GSH), human gastric juice, erythrocyte lysates, and rat tissue postmitochondrial preparations of liver, suprarenal gland, kidney, stomach, and lung eliminate or

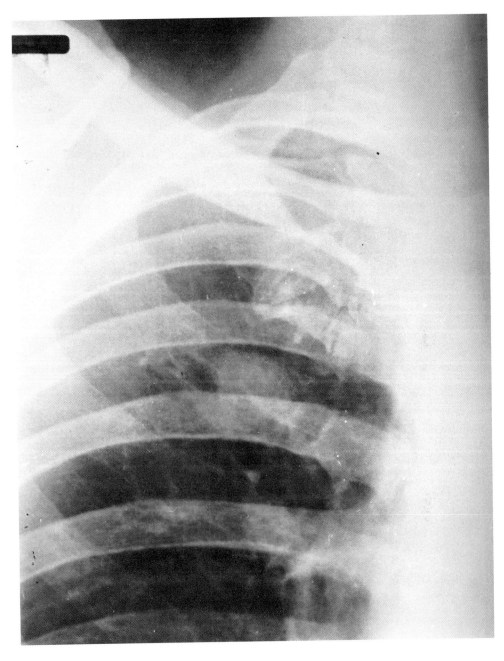

FIGURE 4B

decrease the mutagenicity of hexavalent chromium compounds. Conversely, trivalent chromium compounds become mutagenic only in the presence of a strong oxidizing agent such as potassium permanganate. Postmitochondrial preparations of striated muscle, spleen, colon, and bladder failed to decrease the mutagenic activity of hexavalent chromium compounds. Petrilli and De Flora[21] postulate that deactivation of mutagens by microsomal fractions might be consistent with an intracellular detoxification, hexavalent chromium being reduced to trivalent in the endoplasmic reticulum, and trapped by cytoplasmic ligands. If, however, hexavalent chromium is not reduced in the cy-

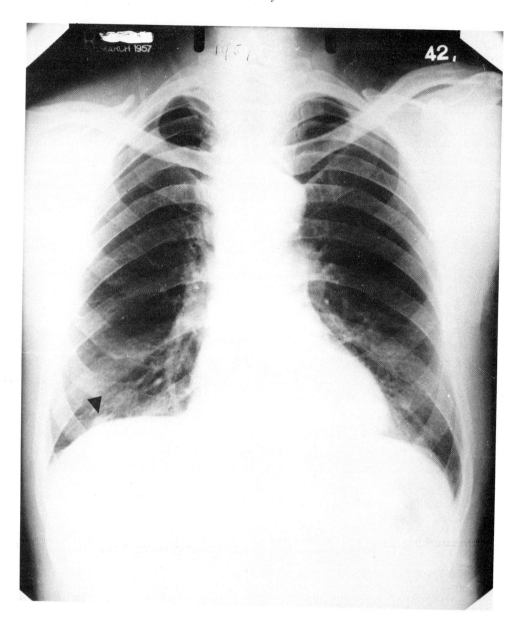

A

FIGURE 5. (A) Shadow right diaphragm, March 1957; (B) same patient, March 1958—adenocarcinoma (Case 7).

toplasm, it may then enter the cell nucleus binding to DNA, as trivalent chromium, and inducing genotoxic effects. Since lung tissue is the least effective of the preparations which decrease the mutagenic activity of hexavalent chromium compounds, the hypothesis proposed by Petrilli and De Flora[21] may explain the observations, consistent in numerous epidemiological studies, that for those exposed to chromium compounds at work, increased risk of cancer is restricted to cancer of the lung. Experience and experimental evidence suggest that slowly soluble compounds may remain longer in

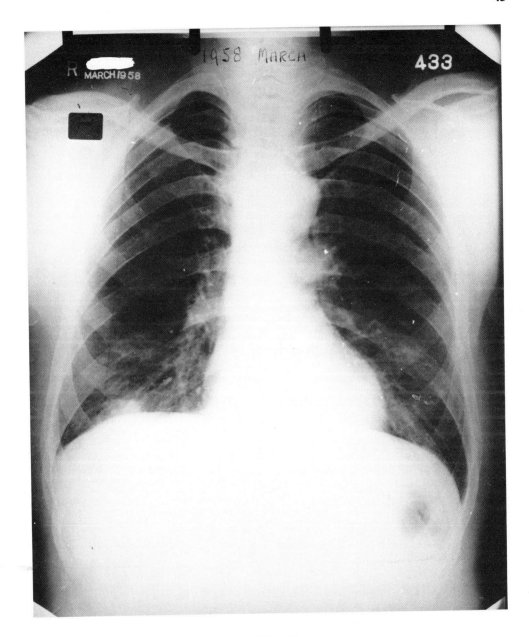

FIGURE 5B

the lung than those which are more soluble, but all are circulated rapidly to other tissues, such as stomach, liver, and kidney, but do not produce tumors at these sites.

VII. INDUCTION OF TUMORS IN EXPERIMENTAL ANIMALS

Early attempts to produce tumors in animals using different species and a wide range of chromium compounds were unsuccessful.[22-24] Attempts to produce bronchogenic carcinoma in mice by exposing them to chromates in inhalation chambers also failed.[25-27] Some increase in pulmonary adenomata was observed.

In 1958, Hueper[28] reported squamous-cell carcinoma coexisting with sarcoma of lung

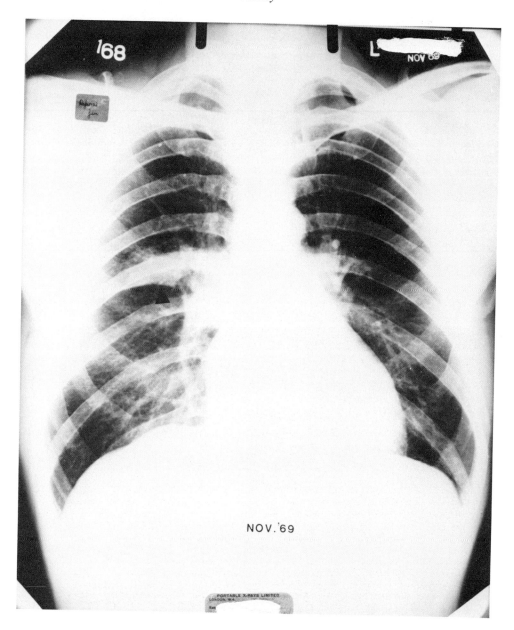

A

FIGURE 6. (A) Right mid-zone shadow, squamous-cell carcinoma; (B) macroradiograph of 6A, survival 5+ years. (Case 8).

and fibrosarcoma of thigh muscle after intrapleural or intramuscular injection of chromite ore roast. This would have contained calcium chromate, a slowly soluble hexavalent compound. Payne[29] and Roe and Carter[30] confirmed that calcium chromate induces injection-site sarcoma.

The relevance of these investigations to the occurrence of lung cancer in men employed in the bichromates-producing industry was difficult to assess but in 1968, Laskin et al.,[31] using the intrabronchial pellet implantation technique developed by Kuschner

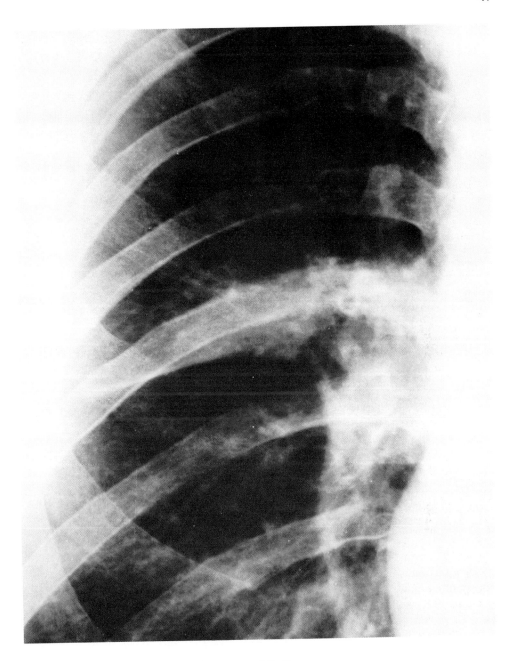

FIGURE 6B

et al.,[32] induced squamous carcinoma in the lungs of rats using calcium chromate and other chromium compounds. The results were statistically significant only for calcium chromate; 8 tumors were found in 100 treated animals. The results were reported in full in 1970.[33]

In 1969, the chromates-producing industry in Great Britain arranged for experiments using the intrabronchial pellet implantation technique to be carried out at Chester Beatty Cancer Research Institute, London, under the direction of the late Professor Sir Alexander Haddow and Professor R. A. M. Case. Fourteen chrome-containing substances were tested, along with appropriate negative (the steel mesh pellet in which the test

Table 1
INTRABRONCHIAL PELLET IMPLANTATION; 100 RATS PER GROUP

Chromates—Series I

Test material	Bronchial carcinoma
Ground chromite ore	0
Bolton high lime residue	0
Residue after alumina precipitation	0
Residue from slurry tank—free of soluble Cr	0
Residue from vanadium filter	0
Residue from slurry disposal tank	0
Sodium dichromate dihydrate	0
Sodium chromate	0
Chromic acid (ground)	1
Chromic oxide (metal)	0
Calcium chromate	8
Chromic chloride hexahydrate	0
Zinc chromate—type II	3
Chrome tan	0
Diphenyl guanidine	0
DPG + calcium chromate	7
Primene 81-R benzoate	0
Primene + calcium chromate	5
Chromic chromate	0
Chromic chromate dispersed in silica	3

From Levy, L. S., Effects of Various Chromium-Containing Materials on Rat Lung Epithelium, Ph.D. thesis, London University, London, 1975.

substance is carried and steel mesh pellet plus cholesterol) and positive controls (3-methyl-cholanthrene at three dose levels). The experiments were done by Dr. L. S. Levy[34] and the results reported in 1975. The results are summarized in Tables 1 and 2.

It will be seen that, as in the experiments of Laskin et al.,[31,33] calcium chromate produced 8 tumors in 100 test animals. This is a statistically significant result. Zinc potassium chromate produced three tumors which is just significant, and chromic acid produced one, which is not. The results in the positive control groups show a clear dose-related response. No tumors occurred in the negative control groups.

These results lend support to the hypothesis, based on experience of lung cancer in the chromates-producing industry, that the likely carcinogenic agent in the process was a slowly soluble hexavalent chromium compound. Further experiments, using the intrabronchial pellet implantation technique on other chromium-containing materials are in progress at Aston University, Birmingham, under the direction of Dr. Levy. Results to date indicate that the carcinogenic potential of hexavalent chromium compounds is related to solubility. Highly soluble compounds such as sodium dichromate do not produce tumors in statistically significant numbers and trivalent and insoluble hexavalent compounds are inactive in this system.

It has been suggested, also, that chromyl chloride, a highly toxic and mutagenic hexavalent compound with a vapor pressure of approximately 20 mm at room temperature, may be carcinogenic and may have contributed to the observed increased incidence of lung cancer in men employed in the chromates-producing industry.[35] The au-

Table 2
INTRABRONCHIAL PELLET IMPLANTATION; 100 RATS PER GROUP

Chromates—Series I

Test material	Bronchial carcinoma
Negative control	
Pellet + cholesterol	0
Blank pellet	0
Pellet + cholesterol + Kieselguhr	0
Positive control	
100% 3-MCA	34
100% 3-MCA	36
50% 3-MCA	18
25% 3-MCA	13
50% 3-MCA	36

From Levy, L. S., Effects of Various Chromium-Containing Materials on Rat Lung Epithelium, Ph.D. thesis, London University, 1975.

thors have shown that this volatile hexavalent compound may be present in the factory atmosphere during manufacture of some chromium compounds, particularly the manufacture of chromic acid if the dichromate used in the process is high in chloride content.

Investigations in the bichromates-producing industry in Great Britain from 1948 onwards have shown that the risks to health are chrome ulceration of the skin or nasal septum; dermatitis and asthma, which are rare; and an increased incidence of lung cancer. There is no evidence of increased risk of cancer at sites other than the lung, nor of excess mortality from causes other than lung cancer. Changes in process, improved working methods and attention to industrial hygiene have reduced, and possibly eliminated, the increased relative risk of lung cancer. Experimental evidence suggests that although all hexavalent chromium compounds, except those which are insoluble, are biologically active, they are not necessarily carcinogenic.

REFERENCES

1. **Korallus, U., Ehrlicher, H., and Wustefeld, E.,** Trivalent chromium compounds—results of a study in occupational medicine, *Arbeitsmed. Sozialmed. Praeventivmed,* 9, 248, 1974.
2. Chrome ulceration of the nasal septum, *Br. Med. J.,* 1, 1364, 1963.
3. **Cumin, W.,** Remarks on the medicinal properties of madar, and on the effects of bichromate of potass on the human body, *Edinburgh Med. Surg. J.,* 28, 295, 1827.
4. **Bécourt, M. M. and Chevallier, A.,** Memoire sur les accidents qui atteignent les oeuvriers qui travaillent le bichromate de potass, *Ann. Hyg. Publique Med. Leg.,* 20, 83, 1863.
5. **Alderson, M. R., Rattan, N. S., and Bidstrup, L.,** Health of workmen in the chromate-producing industry in Britain, *Br. J. Ind. Med.,* 38, 117, 1981.
6. **Machle, W. and Gregorius, F.,** Cancer of the respiratory system in the United States chromate-producing industry, *Public Health Rep.,* 63, 114, 1948.
7. **Baetjer, A. M.,** Pulmonary carcinoma in chromate workers. I. A review of the literature and report of cases, *Arch. Ind. Hyg. Occup. Med.,* 2, 487, 1950.

8. **Bidstrup, P. L. and Case, R. A. M.,** Carcinoma of the lung in workmen in the bichromates-producing industry in Great Britain, *B. J. Ind. Med.,* 13, 260, 1956.
9. **Satoh, K., Torii, K., Fukuda, Y., and Katsuno, N.,** *Proc. Chromates Symposium 80: Focus of a Standard,* Industrial Health Foundation, Pittsburgh, 1980, 61.
10. **Langard, S. and Norseth, T.,** A cohort study of bronchial carconomas in workers producing chromate pigments, *Br. J. Ind. Med.,* 32, 62, 1975.
11. **Davies, J. M.,** Lung cancer mortality of workers making chrome pigments, *Lancet,* 1, 384, 1978.
12. **Davies, J. M.,** Lung cancer mortality of workers in chromate pigment manufacture: an epidemiological survey, *J. Oil Colour Chem. Assoc.,* 62, 157, 1979.
13. **Chiazze, L., Jr. and Wolf, P. H.,** Epidemiology of respiratory cancer and other health effects among workers exposed to chromium. A review of recent literature, *Proc. Chromates Symposium 80: Focus of a Standard,* Industrial Health Foundation, Pittsburgh, 1980, 110.
14. **Hayes, R. B., Lilienfield, A. M., and Snell, L. M.,** Mortality in chromium chemical production workers: a prospective study, *Int. J. Epidemiol.,* 8, 365, 1979.
15. **Hill, W. J. and Ferguson, W. S.,** Statistical analysis of epidemiological data from a chromium chemical manufacturing plant, *J. Occup. Med.,* 21, 103, 1979.
16. **Korallus, U.,** personal communication 1980.
17. **Bidstrup. P. L.,** Carcinoma of the lung in chromate workers, *Br. J. Ind. Med.,* 8, 302, 1951.
18. **Bidstrup. P. L.,** The use of radiology in the early detection of lung cancer as an industrial disease, *Br. J. Radiol.,* 37, 337, 1964.
19. **Bidstrup, P. L.,** Epidemiological survey of the manufacturing (chromates) industry in Great Britain, *Proc. Chromates Symposium 80: Focus of a Standard* Industrial Health Foundation, Pittsburgh, 1980, 100.
20. **De Flora, S.,** Study of 106 organic and inorganic compounds in the *Salmonella*-microsome test, *Carcinogenesis,* 2, 283, 1981.
21. **Petrilli, F. L. and De Flora, S.,** Mutagenicity of chromium compounds, *Proc. Chromates Symposium 80: Focus of a Standard,* Industrial Health Foundation, Pittsburgh, 1980, 76.
22. **Shimkin, M. B. and Leiter, J.,** Induced pulmonary tumours in mice. III. The role of chronic irritation in production of pulmonary tumours in strain A. mice, *J. Natl. Cancer Inst.,* 1, 241, 1940.
23. **Hueper, W. C.,** Experimental studies in metal cancerigenesis. VII. Tissue reactions to parenterally introduced powdered metallic chromium and chromite ore, *J. Natl. Cancer Inst.,* 16, 447, 1955.
24. **Haenszel, W.,** Epidemiological tests of theories on lung cancer aetiology, *Public Health Rep.,* 71, 163, 1956.
25. **Baetjer, A. M., Lowney, J. F., Steffee, C. H., and Budacz, V.,** Effects of chromium on incidence of lung tumours in rats and mice, *AMA Arch. Ind. Health,* 20, 124. 1959.
26. **Steffee, C. H. and Baetjer, A. M.,** Histopathologic effects of chromate chemicals. Report of studies in rabbits, guinea pigs, rats and mice, *Arch. Environ. Health,* 11, 66, 1965.
27. **Nettesheim, P., Hanna, M. G., Jr., Doherty, D. G., Newell, R. F., and Hellman, A.,** Effects of calcium chromate dust, influenza virus and 100R whole-body X radiation on lung tumour incidence in mice, *J. Natl. Cancer Inst.,* 47, 1129, 1971.
28. **Hueper, W. C.,** Experimental studies in Metal cancerigenesis. X. Cancerigenic effects of chromite ore roast deposited in muscle tissue and pleural cavity of rats, *AMA Arch. Ind. Health,* 18, 284, 1958.
29. **Payne, W. W.,** Production of cancers in mice and rats by chromium compounds, *AMA Arch. Ind. Health,* 21, 530, 1960.
30. **Roe, F. J. C. and Carter, R. L.,** Chromium carcinogenesis: calcium chromate as a potent carcinogen for the subcutaneous tissues of the rat, *Br. J. Cancer,* 23, 172, 1969.
31. **Laskin, S., Sohn, L., Teebor, G., and Kuschner, M.,** The experimental induction of lung cancer with chromate compounds, American Industrial Hygiene Association Meeting, St. Louis, 1968 (Abstr).
32. **Kuschner, M., Laskin, S., Christofano, E. E., and Nelson, N.,** Experimental carcinoma of lung, *Proc. 3rd Nat. Cancer. Cnf.,* American Cancer Society, Lippincott, New York, 1957, 485.
33. **Laskin, S., Kuschner, M., and Drew, R. T.,** Studies in pulmonary carcinogenesis, *Inhalation Carcinogenesis,* U.S. Atomic Energy Commission Symp. Ser. No. 18, Washington, D.C., 1970, 321.
34. **Levy, L. S.,** Effects of Various Chromium-Containing Materials on Rat Lung Epithelium, Ph.D. thesis, London University, London, 1975.
35. **Hill, W. H. and Worden, F. X.,** Chromyl chloride—a possibly important industrial air contaminant, *Am. Ind. Hyg. Assoc. J.,* 23, 186, 1962.

Chapter 4

IMMUNOLOGY OF CHROMIUM*

Ladislav Polak

TABLE OF CONTENTS

I.	Origin and Frequency of Allergic Contact Dermatitis to Chromium	52
II.	Significance of Chromium Valency for Sensitization and Testing in Humans	55
III.	Immunological Mechanisms of Contact Sensitivity	57
IV.	Experimental Aspects of Contact Sensitivity to Chromium	61
V.	Immunogenicity and Valency	64
	A. Penetration Capacity of Different Chromium Compounds	64
	B. Binding Capacity of Chromium Compounds to Proteins	67
	C. Conversion of Hexavalent Chromium to Trivalent Forms	69
	D. Immunogenicity of Trivalent and Hexavalent Chromium Compounds in the Guinea Pig	71
	E. In Vitro Reactions Associated with Contact Sensitivity	73
	1. Skin Reactive Factor (SRF)	75
	2. Macrophage Migration Inhibition Test (MMT)	77
	3. Lymphocyte Transformation Test (LTT)	79
	F. Concluding Remarks to the Valency Problem	82
VI.	Genetic Factors in Contact Sensitivity to Metals	84
VII.	Desensitization and Tolerance in Chromium Contact Sensitivity	87
	A. Desensitization	88
	1. Desensitization to Chromium Compounds	89
	2. Effect of Immunosuppressive Agents	94
	a. The Effect of Methotrexate (Table 15, Figure 11)	95
	b. The Effect of Cyclophosphamide (Table 15, Figure 12)	96
	c. The Effect of Antilymphocyte Serum (Table 15, Figure 13)	96
	d. The Effect of a Combination of Cyclophosphamide and Antilymphocyte Serum (Table 15, Figure 14)	96
	e. The Effect of Addition of Prednisolone to the Combination of Cyclophosphamide and Antilymphocyte Serum (Table 15, Figure 15)	98
	B. Induction of Immunological Tolerance to Chromium Compounds	98
	C. Possible Mechanisms of Tolerance and Desensitization	101

*This chapter is an updating of a review published by the same author 1973 in *Progress in Allergy*, Vol. 17, Karger, Basel.

VIII. Secondary Phenomena Occurring during Contact Sensitivity 107
    A. Secondary Flare-Up Reaction 109
    B. Generalized Rash 120

IX. Conclusions 122

References 123

I. ORIGIN AND FREQUENCY OF ALLERGIC CONTACT DERMATITIS TO CHROMIUM

Allergic contact dermatitis is one of the most frequent skin diseases. It can be caused by an almost countless number of substances and theoretically there is no compound in our environment incapable of acting as a sensitizer. However, substantial differences in the frequency of allergic contact dermatitis to these compounds do exist. Some of them become immunogenic only as an extremely rare exception, whereas others are a very common cause of hypersensitivity.

Chromium and its salts belong to the latter group. In extensive clinical studies, it has been shown that chromium is, besides nickel, the most important cause of occupational dermatosis—the frequency of which is continuously increasing with the increasing use of chromium-containing products in our daily life.[1-11] According to some reports, the incidence of allergic chromium dermatitis rose in the 1950s and 1960s by three to four times.[12] Large-scale investigations revealed chromium compounds to be the most common sensitizers in man in Europe.[10,13,14] The average frequency was nearly 11%, ranging according to the industrial character of the countries and regions between 3% (Copenhagen) and 21% (Southern Italy).[15]

The ability of chromium compounds to cause contact dermatitis was first recognized by Parkhurst[16] with the help of patch tests, a diagnostic method introduced by J. Jadassohn[17] some 20 years later. The fact that products containing very small concentrations of chromium salts were the most frequent inducers of skin allergy to chromium, prevented the recognition of the true importance of this sensitizer for a long time. At that time other factors, e.g., alkali content of the incriminated substances,[18] were made responsible for the development of eczema.

However, the discovery by Jäger and Pelloni[19]—that the frequent occurrence of chromium sensitivity among workers in the building industry was due to hypersensitivity to chromium salts present in cement in minute quantities—substantially influenced the clinical research even if allergies to chromium had been described before.[20-23] It has been confirmed that traces of chromium salts form a constant constituent of cements of various origin (Table 1). Chemical properties of cement in respect to its immunogenic activity were reviewed by Fregert and Gruvberger.[24]

Soon after that, it was discovered that small quantities of chromium salts, present in many products used in our daily life,[12,296] are the frequent cause of contact dermatitis occurring among persons having repeated contacts with these products.[25,27] Thus, frequently occurring chromium became, besides nickel, recognized as the most widespread cause of allergic contact dermatitis.

It has been postulated on the basis of experimental and clinical studies that high concentrations of chromium salts are less allergenic than low ones as in the case with

Table 1
ANALYTICAL DATA ABOUT CHROMIUM CONTENT IN CEMENT

Number of investigated samples of cement	% of Cr^{VI} positive	Cr-content in mg %	Ref.
—	100	A few mg/kg	19
67	65	0.1—2.25	14
—	100	2.0—20.0	361
—	—	0.02—2.0	50
32	65	0.14—1.4	362
—	—	ca. 0.05	363
—	80	—	364
22	65	0.2—1.5	83
7	60	0.003—0.7	41
22	10	0.02—0.1	365
10	—	0.1	86
31	100	0.05—2.2	34
45	—	0.02—2.2	166

other haptens.[28,29] In other words, the frequency of allergic eczemas to chromium is higher among people working with material containing small amounts or only traces of chromium salts, e.g., in the building industry, than among people coming into contact with high concentrations of chromium salts, e.g., in industrial manufactures of chromium. In animal experiments, the latter exhibited even an inhibitory effect. Metallic chromium is not immunogenic at all.[30] These views remained, however, not undisputed. The high incidence of chromium dermatitis among chromium platers certainly does not support the mentioned conclusions.[31]

It has been pointed out that, e.g., in cement eczema caused by traces of chromium in this material, some unspecific factors such as the high alkalinity and often occurring damages to the integrity of skin may facilitate the sensitization. Alkaline buffering of chromium salt solutions lowered the threshold concentrations and increased the sensitivity of the skin to epicutaneous challenges with chromium.[14,32-37] Strong acid buffering (pH 1) had a similar effect.[38]

This led some authors to the conclusion that cement eczema is primarily an irritation dermatitis which is fostered and sustained by an additional chromium hypersensitivity.[39-41] The existence of a true contact sensitivity to chromium salts was then questioned by Pautritzel et al.[42,43] These authors suggested that in chromium eczema, chromium itself plays a secondary role in denaturating autologous proteins. This would mean that allergic chromium dermatitis is a sort of autoimmune disease to these "altered self" proteins and that the role of chromium salts consist solely in the transformation of self-proteins to autoantigens. However, this interesting theory was not supported by convincing experimental and clinical data and so was not accepted by the majority of authors. On the contrary, later results disproved Pautritzel's hypothesis.

It has been demonstrated that patients with chromium eczema react not only to the strongly oxidizing potassium dichromate, but also to nonoxidizing chromium compounds such as chromium chloride or chromium alum.[37] On the other hand, strong oxidizers such as potassium permanganate or hydrogen peroxide did not elicit skin reactions in these patients.[44] Chromium tanning liquor used for tanning leather possesses an extremely strong protein denaturating quality which produces very weak skin test reactions in chromium-positive patients.[45] Chromium-containing products, which are

Table 2
THE TEN MOST IMPORTANT EXAMPLES OF ECZEMATOGENIC EXPOSURES TO CHROMIUM COMPOUNDS

Chromium containing materials or objects	Profession or places of contact	Chromium compounds responsible	Ref.
Chromium ore	Industrial chromium production	Chromite	366—368
Chrome baths	Electro-plating industry	Chromic acid Sodium dichromate	369, 370
	Graphic trade	Chromates	148, 112, 371, 372
	Metal industry	Chromates	369
		Zinc chromate	373
Chrome colors and dyes	Painters and decorators, graphic trades, textile, rubber, glass and china industries	Chromic oxide green, chromic hydroxide green, chrome yellow (lead chromate)	21, 22, 366, 373, 374
Lubricating oils and greases	Metal industry	Chromic oxide	369, 375, 376
Anticorrosive agents in water system	Diesel locomotive workshops and sheds, central heating and air-conditioning systems	Alkali dichromates	30, 377—383
Wood preservation (Wolman salts)	Wood impregnation, furniture industry, carpenters, miners	Alkali dichromates	53, 384
Cement, cement products, quick-hardening agents for cement (e.g., Sika 1)	Cement production, manufacture of cement products, building trades	Chromates	19, 34, 385
Cleaning materials (Eau de Javelle), washing and bleaching materials	Housewives, cleaners, laundry workers	Chromates	26, 58, 59, 83, 373
Textiles, furs	Textile and fur industries, everyday life	Chromates	386
Leather and artificial leather tanned with chromium	Leather and footwear industry, everyday life	Chromium sulfate and chromium alum	21, 56, 111, 112, 114, 387—392

the most common cause of chromium allergy, are listed in Table 2.

It is known from several reports that chromium eczema occurs most frequently in workers in the building industry (cement eczema).[46–51] Second in frequency come painters, where the cause is chromium-containing dyes (chromium red, chromium yellow) present in various oil-paints, varnishes, and colored pencils. These dyes are originally water-insoluble and their eczematogenic effect is due to simultaneous use of bases or acids which make the dyes water-soluble.[52] Several other occupations such as galvanizers, machine-drillers, metal-workers, graphic-artists, and employees of the timber, chemical, leather, and textile industries[53–57] are also endangered by the allergenicity of chromium compounds used in these professions. In women, chromium eczema occurs frequently as a consequence of contact with some washing, bleaching, and cleaning materials.[58,59]

In addition, some less frequent causes of hypersensitivity to chromium have been described[60] illustrating how wide-spread this allergy is. To these belong cases of chromium contact dermatitis caused by postage stamps,[61] TV-screens,[62] magnetic tapes,[63] tattooing,[39,64,67] artificial chromium steel-made dentures and bullets, and other chromium-containing foreign bodies retained in the organism[22,60,69,70] and many others reviewed by Burrows[12] and by Polak.[71]

Not very common, but very serious in its consequences, is chromium allergy caused by metallic materials used for internal fixation of fractured bones. This complication very often requires repeated surgery.[70,72]

Another question is whether chromium allergy is not merely a complex hypersensitivity to elements of the VIth group of the periodic system, related to some common effects of these metals. However, hexavalent molybdenum and tungsten salts did not show any relevant effect either in experimental animals (guinea pigs)[73] or in humans.[74] No correlation has been found between immunogenicity of various metals and their position in the periodic system.[75]

The generally accepted view that chromium contact dermatitis is a true contact sensitivity to chromium salts is further supported by the finding that the threshold concentrations of potassium dichromate required for eliciting a positive skin test reaction are extremely low. Geiser et al.[74] described positive results with concentration as low as $10^{-2}\%$, Zelger[37] $10^{-3}\%$, and some other authors even with concentrations between 10^{-4} and $10^{-6}\%$.[32,34,76] These extremely low thresholds point towards the specificity of the chromium allergy being caused by chromium compounds themselves. However, it should be mentioned that the thresholds are largely dependent on the vehicle.[77,78] Another strong proof in this respect is the flare-up reactions of previously positive skin test sites or of healed eczematous lesions upon renewed skin challenges.[79,80] Flare-up reactions also occurred upon i.v. or oral administration of the hapten (potassium dichromate) in hypersensitive individuals.[80,81] All this evidence of chromium eczema being caused by chromium compounds does not, however, exclude the possibility that some secondary influences such as skin irritation by additional chemical (alkalinity, acidity) and mechanical (bruises in building workers) factors may under certain circumstances facilitate the induction or even have a decisive role.[10]

A problem still under discussion is whether the concomitant positive reactions to cobalt and/or nickel sometimes observed in chromium-positive patients are due to cross-sensitivity or to independently, but simultaneously, occurring hypersensitivity to these metals.[74,82–89] With cobalt, this double allergy only occurs in cement eczema where it is known that traces of cobalt are also present in cement. These observations support the view that these patients (according to Zelger[37] 30 to 50% of all chromium-positive ones) were sensitized to cobalt and chromium salt independently. The explanation of the concomitantly occurring positive patch tests to chromium and nickel compounds may be also based on the presence of both these metals in the incriminated products.[90–93] Another possibility of how to explain the simultaneous occurrence of positive patch tests to two noncross-reaction antigens is the so-called "angry back syndrome." This phenomenon, based on the observation that in some contact dermatitis patients positive reactions to some unrelated substances occurred additionally to the reaction to the main antigen when several antigens were applied simultaneously (but not when applied on separate occassions individually), was first described by Mitchell in 1975.[94] The mechanism of this event remains, however, unclear.[95]

II. SIGNIFICANCE OF CHROMIUM VALENCY FOR SENSITIZATION AND TESTING IN HUMANS

Another unsolved problem is the true nature of the hapten which has been a matter of controversy since the work of Jäger and Pelloni.[19] Chromium has six valence states but only two of them, namely tri- and hexavalent salts, are sufficiently stable to be able to form covalent bonds with proteins. This protein-binding capacity is the general precondition of immunogenic activity of hapten shown by Landsteiner in 1935.[96] The question arose as to whether trivalent or hexavalent compounds or both participate in the formation of the final determinant recognized by immunocompetent cells. The following considerations may be of importance. In the process of development of hypersensitivity, two qualities of the hapten are required. The hapten has to possess the capacity of penetrating into the skin and has to be able to form covalent bonds with proteins. Hexavalent salts (potassium dichromate) readily penetrate into the skin but do not form complexes with organic substances. Trivalent salts, on the other hand, form complexes with skin proteins very easily and probably penetrate into the skin with difficulty. This apparent discrepancy was also reflected in the results of patch testing of chromium hypersensitive patients.

Jäger and Pelloni[19] obtained positive results with hexavalent salts (sodium or potassium chromate or dichromate) and consequently concluded that hypersensitivity is directed only against the compounds of hexavalent chromium. These results were confirmed and widely accepted by numerous authors.[21,30,56,67,97–104] However, later on, positive skin reactions to trivalent chromium compounds were also reported by several clinicians.[88,105–115] Some explained their results by contamination of the trivalent chromium compounds used, with traces of hexavalent salts[116,117] or presumed a conversion of the trivalent to hexavalent salts in the skin by oxygen and light. Others were of the opinion that there exist independent hypersensitivities to chromium VI (cement eczema) and to chromium III (leather eczema)[49] so that hypersensitivity is directed against both chromium valencies independently but concomitantly.

In an attempt to clarify the relation of the reactivity of chromium hypersensitive persons to the valency of the compounds used for patch testing, Zelger[37] challenged 700 patients, suffering from chromium eczema, with different concentrations of various chromium compounds (Table 3). Generally, the threshold concentrations of chromium VI compounds were lower than those of chromium III and these, in turn, were lower than those of chromium II.

This has been confirmed by others.[22,118–121] Some authors, in addition to the use of higher concentrations, suggested a longer (48 hr) exposure time.[90,108–110] The frequency of positive patch tests with trivalent chromium chloride was, however, not increased by increasing the concentration of the salt or by using methods enhancing the penetration capacity.[12]

The valency does not, however, seem to be the only factor influencing the results of skin tests in patients with chromium eczema. Bockendahl[105] investigated 101 patients with positive skin tests to hexavalent chromium, 41 of which reacted also to trivalent salts. The conclusion was that the solubility of chromium salts in water was the decisive factor. This view was supported by experiments using lead chromate ($PbCrO_4$), a hexavalent chromium compound which is water-soluble only in limited concentrations.[37] This compound induced positive skin reactions only in patients showing a high degree of hypersensitivity. Less sensitive persons did not react since the threshold concentration required for a positive response was, in these cases, higher than the highest attainable concentration of lead chromate in the solution. Nevertheless, in the highly hypersensitive persons, the threshold concentration of the hexavalent lead chromate capable of eliciting a positive skin reaction corresponded to that of the hexavalent potassium dichromate. Another factor explaining, at least partially, the differences in the results of patch tests with various chromium III compounds (e.g., chloride, acetate, potassium oxalate, or sulfate) is the varying capacity of these compounds to precipitate and form complexes.[90,108–110]

All these results may explain the differences in the results of patch tests performed

Table 3
THRESHOLD CONCENTRATIONS FOR POSITIVE SKIN TESTS IN CHROMIUM-HYPERSENSITIVE PATIENTS

Agent	Valency	Concentration (%)	Number of positive patients
K_2CrO_4	VI	0.001	3
		0.005	9
		0.01	10
		0.05	11
H_2CrO_4	VI	0.001	1
		0.005	5
		0.01	2
		0.05	5
$K_2Cr_2O_7$	VI	0.01	2
		0.05	17
		0.1	14
$KCr(SO_4)_2$	III	0.05	3
		0.1	7
		0.5	12
		Negative	6
$Cr(NO_3)_3$	III	0.05	3
		0.1	2
		0.5	5
		Negative	18
$Cr(CH_3COO)_2$	II	0.1	1
		0.5	3
		Negative	9

Data from Zelger, J., *Arch. Klin. Exp. Dermatol.*, 218, 499, 1964.

with chromium compounds of different valencies but did not answer the question—whether the contact sensitivity is directed exclusively against chromium VI, against chromium III determinants, or both. Under favorable conditions such as a high degree of hypersensitivity, appropriate solvents, sufficiently long exposure time, etc. a positive skin reaction to all chromium compounds used was observed even if different concentrations of them were necessary. Thus, one may agree with Zelger's[37] conclusion, that at least from a clinical point of view, the chromium hypersensitivity is not directed against chromium compounds of a particular valency but rather against the chromium ion as such. Therefore, the more precise term to be used is chromium allergy (hypersensitivity, dermatitis, or eczema) instead of chromate allergy.

Since in human studies, the structure of the chromium determinant inducing hypersensitivity could not be elucidated, a final answer was expected from animal experiments where both sensitization and elicitation with different chromium compounds can be studied in vivo and in vitro.

III. IMMUNOLOGICAL MECHANISMS OF CONTACT SENSITIVITY

For a better understanding of the pathogenesis of allergic contact eczema to chromium it might be useful to present a brief survey of the general immunological mechanism of contact sensitivity.[122]

The research in this field was assisted by the development of a suitable animal model about 50 years ago.[123,124] It has been shown that contact sensitivity induced in guinea

Table 4
MAIN DIFFERENCES BETWEEN CELL-MEDIATED AND ANTIBODY-MEDIATED SKIN REACTIONS

	Contact sensitivity (cell-mediated)	Arthus phenomenon (antibody-mediated)
First appearance of hypersensitivity	5–7 days	14 days
Peak of skin reaction	24–48 hr	2–4
Macroscopic changes	Erythema, induration	Erythema, edema, hemorrhages
Cellular infiltration	Mononuclear cells (macrophages, lymphocytes, basophils)	Polymorphonuclear cells
Transferable with	Lymphocytes	Serum (antibodies)
Proliferation sites in the lymph nodes	Paracortical area	Germinal centers, medullary cords
Type of lymphocytes involved	T lymphocytes	B lymphocytes

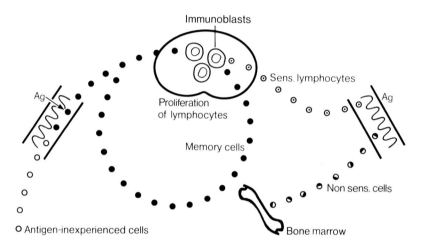

FIGURE 1. Schema of the immune response in contact sensitivity. Hapten penetrates into the skin, where it is recognized by antigen inexperienced specific lymphocytes. These cells migrate to the draining lymph node, where they proliferate and differentiate into memory and effector cells. Sensitized lymphocytes, after leaving the lymph node, circulate in the body. A new contact with the specific hapten elicits an inflammatory skin reaction with the help of nonsensitized cells, secondarily involved in the eliciting phase.

pigs by painting the skin with an allergenic chemical such as dinitrochlorobenzene or picryl chloride[125] closely resembles the allergic eczema in man. Our present knowledge is primarily based on studies using this experimental model. Important immunological data have also been obtained from the recently introduced model of contact sensitivity in mice.[126] However, its relevance for human eczema is disputed.[127]

From an immunological point of view, contact sensitivity is defined as an immunological reaction mediated by a particular type of lymphocytes (T-cells) without a noteworthy involvement of antibodies. The main differences between allergic reactions mediated by antibodies or by other lymphocyte products are listed in Table 4. Operationally, the inductive and the eliciting phases of contact sensitivity may be distinguished (Figure 1).

The inductive phase is initiated by biochemical and biological processes leading to

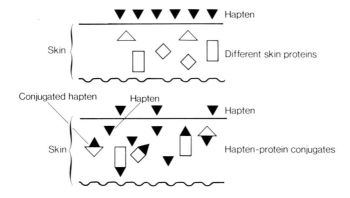

FIGURE 2. Formation of the antigen. The hapten applied to the skin penetrates into the skin, forming immunogenic conjugates with different autologous proteins (preparatory phase).

the formation of an antigenic complex (preparative phase). It is generally accepted that substances of low molecular weight, called haptens, which are the usual cause of allergic skin reactions, become immunogenic only when covalently bound to a protein carrier (Figure 2). In contact sensitivity, this carrier was looked for among the proteins of the skin and the serum, since these are the first which the hapten encounters when penetrating into the skin. However, the true nature of this carrier remained a controversial matter for a long time. It has been assumed that the haptens bind *in situ* several proteins, thus forming a set of multiple conjugates.

The recent discovery that macrophages or macrophage-like cells (Langerhans' cells) are indispensable for in vitro stimulations of T-lymphocytes helped to clarify the problem of the formation of the antigenic complex in contact sensitivity.[128] It has been shown that guinea pigs injected intradermally 14 days later with hapten-modified macrophages exhibited a clearly positive skin reaction to the epicutaneously applied hapten, providing that macrophages from syngeneic donors were used for sensitization. Hapten-modified allogeneic macrophages, however, completely failed in this respect.[128a] In further experiments, it has been shown that this genetic restriction is on the level of antigen presentation.

Delayed-type hypersensitivity reactions are conditioned by the homology not only of the hapten but also of the carrier (carrier specificity).[129,130,289] The failure of haptenized allogeneic macrophages to induce contact sensitivity may therefore be considered as evidence that a genetically controlled macrophage moiety is a decisive component of the antigenic complex. Since the antigenic complex formed by haptenization of allogeneic macrophages differs from the complex formed by the epicutaneous application of the hapten in animals injected with these macrophages in respect of this moiety, the skin reaction cannot be elicited.

Recently, it has been found that dendritic cells in the epidermis, called Langerhans' cells, are particularly capable of participating in the formation of the antigenic complex (Figure 3). These macrophage-like cells are important for both induction and elicitation of contact sensitivity, since they are the first cells of this type encountered by the hapten penetrating the skin. Moreover, they are present also in the lymph nodes, suggesting that the antigen recognition may take place both in the skin and in the draining lymph node.[131]

An early (up to 12 to 18 hr) excision of the hapten application site prevents the induction of contact sensitivity.[132] The excised skin sample grafted on an isologous

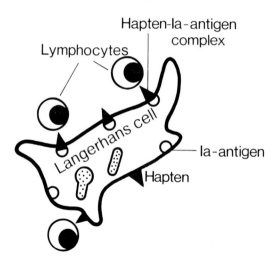

FIGURE 3. Recognition of the antigenic complex by specific lymphocytes. Hapten or hapten-protein conjugates are first processed by the macrophage-like Langerhans' cells expressing the Ia-Antigen. The immunogenic structure is a complex of the hapten with Ia-antigen probably under participation of some other autologous skin proteins.

recipient induced contact sensitivity only when excised not later than 3 hr after hapten application.[122] These results indicate that both the proportion of the hapten remaining in the skin and the proportion of the hapten absorbed into the circulation participate in the induction of contact sensitivity. Nevertheless, intact lymphatic connections between the skin area where the hapten was applied and the draining lymph node are indispensable for the induction of contact sensitivity.[132]

Lymphocytes leave the skin via lymphatics and are (in sensitized animals) retained in the paracortical area of the draining lymph node where they proliferate and differentiate into specific memory and effector cells. The peak of proliferation was reached on day 4 after the application of the hapten for sensitizing purposes.[133] At that time, a maximum of immunoblasts was observed in the draining lymph node. On day 5, the number of immunoblasts started to decrease and the newly developed effector and memory cells were leaving the lymph node and were dispersed in the organisms mainly via the blood vessels.

The proliferative response of lymphocytes from sensitized animals can be demonstrated by an increased DNA-synthesis upon antigenic stimulation in vitro, the so-called lymphocyte transformation test (LTT).[134,135]

A removal of the draining lymph node before memory and effector cells are formed, prevented or considerably diminished the degree of contact sensitivity.[136] On the other hand, lymphocytes from the draining lymph node taken at the peak of proliferation and differentiation are capable of transferring contact sensitivity to syngeneic recipients.

The state of hypersensitivity becomes apparent when the animals are challenged epicutaneously with the specific hapten (eliciting phase). This second contact with the hapten initiates a sequence of events ending in an inflammatory skin reaction and at the same time, in an increase in the degree of contact sensitivity (booster effect) due to stimulation of memory cells. Specific effector cells present at, or attracted to, the sites of hapten application are stimulated by the antigenic complex formed in the skin.

They respond with a release of mediators known as lymphokines.[137] These substances mobilize secondary mononuclear cells which amplify the inflammation-inducing effect of the relatively small, and therefore insufficient, proportion of specific effector cells.

This is not, however, the only effector mechanism in contact sensitivity. An almost complete depletion of white blood cells by irradiation or cytotoxic treatment shortly before an epicutaneous challenge of contact-sensitive guinea pigs did not substantially change the resulting macroscopic reaction. The infiltrate was, however, almost completely suppressed. It is evident that contact sensitivity reactions, occurring in apparent absence of the amplifying mechanism, are induced by mediators other than classical lymphokines and originate from cells or tissues resistant to irradiation or cytotoxic drugs. It seems that the eliciting mechanism in contact sensitivity is a complex one, and that various mediators can replace each other to secure the functioning of these originally defense mechanisms.[138]

Recently, it has been discovered that allergic reactions are controlled by intrinsic control mechanisms, namely suppressor cells. These cells are activated by the same procedures which activate memory and effector cells, are of the same specificity, and are capable of limiting and, in extreme cases, of preventing the development of allergic reactions.[139]

Depending on the mode of induction, they affect the afferent, central, or efferent limb of the immune response.[140]

The evidence for the existence of suppressor cells in contact sensitivity rests on three findings:

1. Some cytotoxic agents, such as cyclophosphamide, known to eliminate suppressor cells, enhance the resulting degree of contact sensitivity when applied shortly before the sensitizing procedure.[141]
2. Transfer of lymphocyte suspension from tolerant or normal donors containing suppressor cells or their precursors abolished this enhancement and restored the normal degree of contact sensitivity.[142]
3. Extreme activation of suppressor cells is achieved by an i.v. application of a high dose of the hapten. This procedure induces a complete and permanent specific unresponsiveness (tolerance).[143] This tolerance is transferable to normal recipients by parabiosis and reversed by elimination of suppressor cells by cyclophosphamide treatment.[144]

The exact mechanism by which these suppressor cells control the development and intensity of skin reactions, however, is not yet fully understood. It has been suggested that a specific suppressor factor produced by suppressor cells is responsible for this control mechanism.

IV. EXPERIMENTAL ASPECTS OF CONTACT SENSITIVITY TO CHROMIUM

A development of a suitable animal model for contact sensitivity to chromium is a precondition for obtaining more information about the pathogenesis of chromium allergic contact dermatitis. Since guinea pigs cannot become sensitized by epicutaneous application of chromium salts in aqueous solutions,[49,145–148] it was necessary to find more appropriate solvents. It is a clinical experience that some detergents and alkali, beside producing mechanical lesions of the skin (e.g., microtraumata due to cement), play an important role in developing this allergy and animal experiments confirmed this view. Contact sensitivity to chromium was reported to be induced by painting the skin with chromium salt dissolved in sodium lauryl sulfate,[149-153] Triton® X-100,[73,154] or dimethylsulfoxide (DMSO).[147] Others have failed to confirm this.[155]

Some authors were successful in inducing chromium contact sensitivity by enhancing the skin permeability by pretreatment of the hapten application site with petrol or heparin.[156] However, the best results in inducing contact sensitivity to chromium were achieved by injecting chromium salts intradermally homogenized in Freund's complete adjuvant (FCA),[28,157-159] but the results were not positive in all animals. Thus Magnusson and Kligman[145] induced chromium contact sensitivity in 21 of 25 guinea pigs with the help of FCA, in 10 of 25 with the help of Freund's incomplete adjuvant, and in only 1 of 25 without the use of adjuvant.

In order to improve the proportion of sensitized animals and the degree of achieved contact sensitivity, a combined method consisting of several intradermal and epicutaneous applications has been elaborated using potassium dichromate as hapten.[160] Guinea pigs weighing at least 400 g and maximally 600 g received five injections of 0.2 mg potassium dichromate in 0.2 mℓ FCA each intramuscularly into the footpads or thighs and the nape of the neck. The total dose per animal was 1 mg. Two weeks later the animals were boosted by an intradermal injection of 25 µg potassium dichromate in 0.1 mℓ 0.15 M NaCl into the clipped skin of a flank. At the same time, the clipped skin of the other flank was painted with 0.025 mℓ of a 0.5% solution of potassium dichromate in 1% Triton® X-100. The epicutaneous applications were repeated every week using the same application site marked with a stamp and the intradermal injections were repeated every 14 days. The reactions to the epicutaneous challenges were read 24 hr later. A positive skin reaction appeared usually after two to three boosters, i.e., after 6 to 8 weeks as a confluent redness sometimes combined with swelling. At that time, a simultaneous epicutaneous challenge of an untreated skin site also gave a positive, though weaker, reaction. The reactions were graded 2+ for bright red and confluent and 1+ for faint, pink, and not confluent. The reactions lasted 48 to 72 hr and dissolved with scaling of different intensity during the following days.

Because the results obtained with different sensitizing methods vary considerably, it appeared to be of importance to compare the sensitizing capacity of various methods using the same methods for challenging and the same stock of guinea pigs. This task was undertaken by Skog and Wahlberg[161] and by Hicks et al.[162]

Skog and Wahlberg sensitized albino guinea pigs, at average weight of about 300 g at the beginning of the experiment, by the methods developed by Hunziker,[158] by Magnusson and Kligman,[145] and by Polak and Turk.[160] All these methods have in common the use of potassium dichromate in FCA for the first injection followed by a second injection of the hapten in FCA on the following day (Magnusson and Kligman) or a week later (Hunziker). The Polak and Turk method has been described in detail above. The efficiency of these methods was established with various challenging techniques such as patch testing, intradermal challenge, and application of the hapten in ointment as well as by systemic transfer of chromium contact sensitivity with blood cells from hypersensitive guinea pigs.[168] Histological picture evaluating acanthosis, edema, and cellular infiltrate was used as an additional parameter. All three sensitizing methods induced contact sensitivity to chromium, the Polak and Turk method being somewhat better but more time-consuming.

Hicks et al.[162] compared the following sensitizing techniques: guinea pig maximization test,[145] split adjuvant technique,[163] and the methods of Polak and Turk[160] and of Gross et al.[157] Again, all these methods used potassium dichromate in FCA. Contact sensitivity to chromium salts was induced by all these methods; however, the results were somehow better with the guinea-pig maximization test and with the Polak and Turk method than with the other two.

Table 5
TIME OF ONSET OF Cr-HYPERSENSITIVITY

Time after sensitization	Number of sensitive animals	% of sensitive animals
2 weeks	24	35
4 weeks	9	13
6–8 weeks	23	33
Later	2	3
Nonsensitized	11	16

From the survey of sensitizing techniques, it is evident that reproducible results with a high frequency of positive animals were obtained only when FCA was used. However, there exist important differences between contact sensitivity induced with or without use of FCA.[135] A reliable sensitizing technique without use of adjuvants would be more related to practice and might supply important additional informations about the pathogenesis of contact sensitivity to chromium as is known from the clinic.

Recently, high degrees of contact sensitivity to dinitrochlorobenzene, picryl chloride, and oxazolon were achieved by using hapten-modified syngeneic macrophages injected intradermally without FCA. This method is based on the knowledge of the role of macrophages and macrophage-like cells such as Langerhans' cells in the formation of the antigenic complex and in its recognition. This method was also used for induction of contact sensitivity to chromium with partial success.

The methods of challenging are also important, but the sensitivity, specificity, and reproducibility of various methods are disputed. Some authors prefer the intradermal route of challenges,[28,164] others considered the epicutaneous application of the hapten dissolved in water or a detergent as more reliable and certainly more praxis-related.[46,116,165,166]

Skog and Wahlberg,[161] who compared five challenging methods, namely patch test, intradermal injection, application of the hapten in ointment, open epicutaneous painting, and application on the nipples, recommended the intradermal method and the application in ointment as giving the most reliable results. Hicks et al.[162] on the other hand, considered painting and patch testing as the most effective challenging procedures as far as sensitivity and reproducibility were concerned. Polak and Turk[160] found the intradermal injection as the most sensitive method giving the earliest positive results. However, epicutaneous application of chromium salts in detergent (1% Triton® X-100) was most suitable for quantitative evaluation of intense hypersensitivity. The epicutaneous challenge also reflects more closely the clinical conditions under which the elicitation of allergic contact eczema often occurs.

During the elaboration of the sensitizing methods used in the following experiments, we observed that the time interval between the first application of the hapten and the first manifestation of contact sensitivity varied considerably from animal to animal. In order to clarify this problem, we sensitized a larger number of Hartley guinea pigs with our combined method consisting of intramuscular injections of potassium dichromate in FCA, supported by several intradermal and epicutaneous applications of the hapten during the following weeks. The epicutaneous challenges were always evaluated according to the arbitrary scale mentioned above. The results of this experiment are given in Table 5.

One third of the animals (35%) were fast reactors showing strong contact sensitivity to the first epicutaneous challenge performed 14 days after the sensitizing procedure was started. The slow reactors (33%) exhibited positive skin reactions to challenges performed 6 to 8 weeks after the sensitization had begun. About 15% remained resistant due either to natural (genetically determined) tolerance or to tolerance acquired during the process of sensitization.

V. IMMUNOGENICITY AND VALENCY

The proportion of sensitized animals also showed variations related to seasonal, dietary, and other factors but this was not followed up in our studies. Such influences are, however, well-known from the literature.[167,250,264] Having established a reliable and efficient method for inducing contact sensitivity to chromium salts, we tried to investigate the relation between sensitizing capacity and valency. In the experimental animal model the following questions were studied.

1. The relation between the skin penetration capacity of different chromium salts and their valency. The penetration of the hapten into the skin and its retention there are important preconditions for allergenicity of the haptens.
2. The relations between protein binding capacity of different chromium salts and their valency. Since Landsteiner's work,[96] this parameter continuously gained in importance and was recently completed by the finding that the most important proteins in this respect are present on macrophages and Langerhans' cells.[169]
3. The possible conversion of chromium compounds of a certain valency into compounds of another valency that might be more suitable to form antigenic complexes of higher sensitizing or eliciting capacity.

A. Penetration Capacity of Different Chromium Compounds

The penetration of chromium compounds into the skin and their disappearance from the skin was studied either in vivo or in vitro with the help of excised skin samples. Different species such as guinea pigs, hairless rats, mini pigs, and for in vitro experiments human skin samples were used also for these studies. The chromium salts were either labeled with radioactive chromium or the content of chromium was evaluated by colorimetric, spectrographic, or polarographic techniques.[170,171]

In in vivo experiments, chromium salts were applied epicutaneously or intradermally, or the iontophoretic technique was used. The penetration was determined by measuring the chromium content in the blood, urine, and other organs and tissues.[172–177] Important information was also obtained by studying the content and the progressive disappearance of the metal from the skin samples excised at different times after its epicutaneous application. In in vitro experiments, excised skin samples or single skin layers were placed in special diffusion chambers and the amount of chromium, penetrating through or retained in the skin, was evaluated by different methods. A less accurate method consisted in an estimation of chromium in pieces of skin dipped for a certain time into chromium solutions.

In numerous papers, the route of penetration of chromium compounds into and through the skin and its dependence on the concentration and pH of the applied solutions was studied. It is known that substances from the environment enter the skin either via the epidermis or via epidermal adnexa. It is of interest that the deeper layer of the "devitalized" stratum corneum exhibited the highest resistance to penetrating substances.[173–175] When this barrier is surmounted, the rest of the epidermis does not substantially impede the diffuse penetration. This view was, however, considered as a result of a mathematical error by Blank.[177a] According to Blank the barrier is evenly spread over the whole stratum corneum. Sometimes, the membrana basalis forms a second barrier which may retain the proportion of the hapten important for sensitization. The passage of substances applied onto the skin can also proceed through the skin adnexa. Important in this respect are the hair follicles and the sebaceous glands, whereas sweat glands seem to play a relatively insignificant role.[175]

Several authors have shown that salts of hexavalent chromium, such as potassium dichromate or sodium chromate in relatively high concentrations (1%) penetrate the skin virtually unimpeded and can be traced in the blood and urine. Chromium compounds at lower concentrations (0.1 to 0.001%) penetrate, however, only into the skin and remain there.[173,178] This formation of antigenic depots may constitute a prerequisite for sensitization and explain why small amounts of low concentrated hexavalent chromium compounds present, e.g., in cement, are a most frequent cause of eczema whereas contact with highly concentrated chromium compounds occurring, e.g., in metal industry, leads less often to this allergic disease. There is, however, another additional explanation. It is known that the hapten which is absorbed into the circulation, probably in an unconjugated form, may induce immunological tolerance.[29] Since the amount of the hapten reaching the blood stream is usually proportional to the amount applied to the skin, absorption of higher amounts of hapten may prevent the development of allergic eczema even if sufficient amounts are retained in the skin in an allergenic form. This phenomenon, described as tolerance during the primary response[179] will be discussed later on.

The dependence of the penetration of the hexavalent chromium salts on their concentration in the solution was extensively studied by Wahlberg.[180] He elaborated an isotope technique called disappearance measurements where the radioactivity above a skin depot of the labeled ions was continuously recorded for 5 to 25 hr with a scintillation recorder. In guinea pigs, the maximal relative absorption occurred at a concentration of 0.261 N sodium chromate and was 4% of the applied amount. Higher and lower concentrations gave smaller absorption values (1%). The maximal absolute penetration was reached at the concentration of 0.398 N. This "plateau value" (which cannot be further increased) corresponds to 700 mM sodium chromate per hour per square centimeter.[176,181] However, when the skin samples were excised and the disappearance of chromium measured in vitro, the penetration of human and guinea pig skin was proportional to the concentrations used and was in contrast to in vivo experiments not limited by a "plateau value." Generally the penetration was greater in the guinea pig than in man.[182]

The disadvantage of the disappearance measurement method is that it expresses only the penetration through the whole skin without being able to distinguish between the proportion penetrating the surface layers but retained in the deeper layer of the skin and the proportion penetrating the whole skin and disappearing in the circulation since both are recorded by this technique. This discrimination is important for assessing the allergenicity of the investigated materials since only the proportion retained in the skin is immunogenic. Furthermore, it has been demonstrated that the penetration of chromium compounds through the skin is pH-dependent. It is higher at alkaline values than at acid ones and at the same pH is higher with buffered solution than with unbuffered ones.[183] The method of application of the investigated compounds also plays an important role. Experimental results are however controversial and available data too scarce to allow a definite conclusion.

Some authors found the penetration of sodium chromate after iontophoretic application to be up to 40 times greater than after the epicutaneous one.[184,185] This difference was more pronounced at lower concentrations than at higher ones. On the other hand, it has been reported that the resorption from the skin after iontophoretic application was delayed in comparison to the intradermal administration.[186]

Differences in the absorption of hexavalent chromium salts from the skin of normal and chromium hypersensitive individuals were never clearly demonstrated. According to Czernilewski,[172] the absorption in sensitized animals is lower than in normal ones. On the other hand, Heise and Matthäus[187] observed an enhanced penetration into the eczematous skin areas whereas van Kooten et al.[170] failed to find any difference in the chromium content in the skin of eczematous and noneczematous patients.

In a detailed study, Pederson et al.[188,189,193] examined the content of radioactive sodium chromate in the skin of normal and contact-sensitive persons after intradermal or epicutaneous (patch test) applications. No difference was observed between hypersensitive and normal individuals after intradermal application; 50% of the applied dose disappeared within 10 min, 14 to 18% was present at the application site after 2 days, and 3 to 5% was still traceable after 2 months. Injected chromium was further found in blood cells and plasma. Epicutaneous application (patch test) of 20.8 µg of radioactive sodium chromate gave a quite different picture. In normal persons, 7.8 to 10.3% of the applied amount was detected in the skin after removal of the patch test. One month later 0.5 to 1% of the dose used still remained at the hapten application site (0.13 to 0.22 µg). This is in agreement with a previous report that small amounts of metallic salts remained for a relatively long period of time on the surface of the skin.[190-192] However, in hypersensitive patients the detected amount of chromium was significantly lower and in about 3 weeks completely disappeared from the skin test site. The apparent discrepancy of the results obtained after intradermal or epicutaneous application of the hapten in allergic subjects could be explained by the fact that chromium in the epidermis is shed together with epidermal cells by the inflammatory reactions elicited by the hapten. This inflammation, however, does not occur in normal (not hypersensitive) hapten-treated persons. Chromium in the dermis, on the other hand, is bound and only slowly eliminated. After epicutaneous application, no chromium could be detected either in the blood cells or in plasma.

From all the experiments discussed above, one conclusion can be drawn despite different methods used and different interpretations of their significance. There is no doubt that salts of hexavalent chromium are capable of penetrating the skin quite easily. This is, however, not as clearly demonstrated for salts of trivalent chromium. This problem seems to be more complicated and the reports in the literature are rather contradictory. In the studies of Samitz and Gross[56] and Schwarz and Spier,[174,175] no differences in the penetrating capacity of hexavalent and trivalent chromium compounds were detected. On the other hand, Mali et al.[194] reported that only very small amounts of hexavalent chromium are capable of penetrating into the human skin in vivo and into the guinea pig skin in vitro. The penetrating capacity of trivalent chromium into human skin in vitro is, according to Spruit and van Neer,[195] 10^{-4} times smaller than the capacity of hexavalent salts. It has been shown that different trivalent chromium salts have different penetrating capacities[196,197] when used in equivalent concentrations. The undamaged human skin is not penetrated by chromium III sulfate and only marginally by chromium III nitrate, whereas chromium III chloride penetrates almost as well as the hexavalent potassium dichromate. When the stratum corneum had previously been removed (stripped skin), all chromium III salts penetrate the skin equally well. These results explain the positive skin reactions on stripped skin, obtained with various chromium III compounds. Experiments on isolated human skin in vitro gave similar results.[198]

Two conclusions can be drawn from these experiments:

1. The penetration of various compounds of trivalent chromium (chloride, sulfate, nitrate) into the skin depends largely on their ionic strength and pH.
2. The difference in the penetration capacity between different chromium III com-

pounds and between chromium III and chromium VI salts is due to the epidermal layer that forms the main barrier.

Wahlberg and Skog[177] also compared the penetration capacity of trivalent and hexavalent chromium compounds using their disappearance measurement method. They found that chromium salts of both valencies applied at low concentrations (0.017 to 0.239 M) penetrate the skin equally well. At higher concentrations (0.261 to 0.398 M), the penetration of the hexavalent sodium chromate is much faster than that of the trivalent chromium chloride. The "plateau value" of chromium chloride was about 50% lower (300 nmol/hr/cm^2 than for sodium chromate. These results apply only to in vivo conditions with undamaged skin. In vitro, no differences in the penetration capacity between trivalent and hexavalent chromium salts were observed.[185] The penetration through the living skin therefore seems to be limited not only by mechanical but also by biological factors.

The general conclusion to be drawn from all experiments dealing with penetration of chromium of different valencies into or through the skin is, that under the same conditions hexavalent salts penetrate more easily and faster than trivalent ones. For a correct interpretation of these results several additional factors such as the chemical structure and the concentration of the individual compound as well as the conditions of the skin, the types of assay (in vivo or in vitro), and the physicochemical properties of the solvent must be considered.

B. Binding Capacity of Chromium Compounds to Proteins

From an immunological point of view it is important to know what proportion is retained in the epidermis and bound to Langerhans' cells or macrophages and what proportion is absorbed into the circulation. The first proportion is considered as immunogenic, the latter as tolerogenic.

The penetration through or retention in the skin are, however, not the only decisive factors in the sensitizing process. If the amount of hapten is sufficiently high and the resorption fast, tolerance may occur even in the case when enough allergenic material is formed and bound in the epidermis (tolerance during the primary response). More information about the relation between chromium compounds of different valencies and the macrophages or Langerhans' cells is necessary before the exact importance of the penetration capacity for the sensitizing process can be established.

Substances of low molecular weight, to which chromium compounds belong, become immunogenic only when conjugated with protein carriers.[96] Therefore, the investigation of the differences in the protein-binding capacity between trivalent and hexavalent chromium salts may help to clarify the role of valencies for the sensitization. Numerous in vivo and in vitro experiments have demonstrated that only trivalent chromium compounds are capable of conjugation with proteins of the skin and other organs and of the blood. Hexavalent chromium salts, on the other hand, are taken up by cells, i.e., they penetrate the cell membranes (e.g., erythrocytes). This difference may be of great importance in view of the decisive role of Langerhans' cells and macrophages in the sensitizing process. However, little data were available up to now on this particular subject. Apposition of lymphocytes to Langerhans' cells has been found only sporadically in electron microscopic pictures from chromium positive human patch tests.[199] Shelley and Juhlin have demonstrated that chromium salts could be taken up by Langerhans' cells.[169]

Nevertheless, a brief survey of the literature about the distribution of chromium compounds in the body after their parenteral application and their interaction with proteins and cells appeared to be of importance for assessing the role of these haptens in contact sensitivity. In guinea pigs, after an i.v. injection of radiolabeled trivalent chromium chloride, 94 to 99% is bound to plasma proteins in a stable form. Similarly labeled hexavalent sodium chromate, on the other hand, is taken up by erythrocytes.[200,201] The results after intraperitoneal or intratracheal applications were not different.[202] Furthermore, it has been reported that hexavalent chromium salts form, if at all, very unstable conjugates, whereas trivalent chromium compounds are unable to penetrate cell membranes.[203]

When assessing the binding capacity of trivalent chromium to different serum proteins, the largest amounts were detected to be conjugated with albumins followed by α-, β-, and γ-globulins and fibrinogens.[204] However, it has been reported that very small amounts (traces) of trivalent chromium are bound predominantly to transferrin (siderophillin) and that only larger amounts form conjugates with other plasma proteins, preferentially with β-globulins.[68,205,206] Anderson[210] compared the binding capacity of hexavalent and trivalent chromium to skin proteins and found trivalent salts to be more active in this respect.

The good binding capacity of trivalent chromium to various proteins and the marginal or even lacking capacity of hexavalent chromium compounds was confirmed by numerous authors.[73,174,194,196,208,209] In the rare cases where hexavalent chromium was found to be bound to proteins, it has never been established whether the conjugated chromium was still present in its hexavalent form or converted into the trivalent one.[210] As to the localization of chromium conjugates in the body, it has been found that after i.v. injection of trivalent chromium compounds into rats, the highest concentrations of protein-bound chromium were found in the bone marrow and somewhat smaller amounts in the spleen and liver,[211] 40% of the injected amount was excreted via the kidneys. After epicutaneous application of chromium compounds, it has been found that lymph nodes contains more chromium than spleen or liver.[174]

In order to confirm data from the literature, the author performed the following in vitro experiments. Fresh guinea-pig serum (25 mℓ) was incubated with radiolabeled chromium chloride ($^{51}CrCl_3 \times 6H_2O$; 60 mg) for 1 hr at 37°C and then dialyzed against 0.15 M NaCl for 20 hr at 4°C to remove nonconjugated hapten. Afterwards, the serum was centrifuged for 30 min at 3000 rev/min and the chromium content estimated by measuring the radioactivity in a γ-counter. In 1 mℓ serum, 1.5 mg chromium chloride was detected. However, when guinea-pig serum was incubated under the same conditions with hexavalent (potassium dichromate) instead of trivalent chromium compounds, no measurable amounts of the hapten bound to proteins were detected. The content of potassium dichromate was measured using diphenyl-carbazide reagent.[210-212] By separation of chromium-conjugated serum proteins on a Sephadex® G-200 column with phosphate-buffered 0.15 M NaCl (pH 7.2), chromium was found mainly in the albumin fraction—1 mg protein contained 8.8 μg elementary chromium.

The binding capacity of trivalent chromium salts to proteins was further investigated by Samitz and Katz.[208] According to the authors, the maximal binding capacity of trivalent chromium salts to soluble skin proteins was 0.106 mg elementary chromium to 1 g proteins. Also, carboxyl groups and not sulfhydryl groups are mainly responsible for this process. Their conclusion was that trivalent chromium is attracted to proteins as cation whereas the hexavalent is repelled as anion. Magnus[213] also failed to demonstrate the binding capacity of hexavalent chromium to cyanide-keratin, sulfide-keratin, insulin, or histon. In his experiments, paper electrophoresis at pH 2 to 10 was used. The findings of Sertoli and Panconesi[171] (who were unable to detect any binding of trivalent chromium salts to amino acids, peptides, and soluble proteins) are somehow isolated and unexplainable. For detection of chromium, chromatography and autoradiography were used. Their negative results with hexavalent chromium have merely augmented the already quoted data.

The immunological value of conjugates of trivalent chromium remained, however, unclear. The majority of authors were unable to sensitize animals with in vitro prepared conjugates of trivalent chromium salts with albumins, globulins, or skin extracts. Moreover, sensitized animals or humans failed to show any positive skin reaction to challenges with these conjugates.[157,209]

There are only two exceptions mentioned in the literature. Cohen[106,107] obtained positive skin reactions in potassium dichromate hypersensitive patients to intradermal challenges with conjugates of chromium chloride to heparin, human serum albumin, and serum γ-globulin; with the last one, however, only in 30% of the patients tested. An interesting result was reported by Samitz et al.[197] These authors described positive skin reactions in chromium-hypersensitive guinea pigs challenged intradermally with a chromium-cystin conjugate. Hypersensitive humans failed, however, to react to this conjugate. Later on, Katz et al.[214] attempted to elicit positive skin reactions in chromium hypersensitive guinea pigs by intradermal injections of various conjugates of chromium chloride with guinea pig serum albumin, guinea pig gamma globulin, heparin, or chondroitin. The chromium albumin complex proved to be the only efficient elicitor.

From the data reviewed in the two preceding sections, the discrepancy between two important preconditions of immunogenicity of haptens, namely the capacity of penetrating into the skin and the capacity of forming covalent bonds with proteins, becomes apparent. Salts of trivalent chromium bind easily to all types of proteins forming stable conjugates but are unable to penetrate the undamaged skin. Salts of hexavalent chromium on the other hand, penetrate easily into the skin but form stable conjugates with proteins only with great difficulties. It is known from clinical experience that hexavalent chromium compounds such as potassium dichromate are much more efficient in inducing and eliciting contact sensitivity reaction than trivalent compounds. Several attempts have been made to explain this discrepancy.

The view of Pautritzel et al.[43] that the actual antigen is a protein denatured by hexavalent chromium, seems to be unlikely because of the negative results obtained with denatured proteins and other haptens having the same protein denaturating effect.[37,197]

Also a suggestion by Everall et al.[45] that chromium (and other heavy metals such as nickel and cobalt) do not require to be bound to protein carriers for becoming immunogenic, is not sufficiently documented by experiments.

The view that antigenic complexes are formed by conjugation of hexavalent chromium with lipids or with other nonprotein substances of high molecular weight remains a nonconfirmed hypothesis.[174]

Therefore, the most likely possibility remains that hexavalent chromium is, after penetrating into the skin, reduced to its trivalent form and only as such bound to proteins in order to become immunogenic. As a support for this hypothesis, it has been shown in numerous experiments, in vivo and in vitro, that hexavalent chromium can indeed be reduced to its trivalent form after entering into contact with various constituents of the skin.[194,196,208,215,216]

C. Conversion of Hexavalent Chromium to Trivalent Forms

Samitz and Katz[208,216] have shown that 1 g of skin possesses the capacity to reduce in vitro 1.06 mg hexavalent potassium dichromate to a trivalent compound. Furthermore, it has been shown that the sulfhydril (SH) groups containing amino acids such as cystine, cysteine, and methionine are mainly responsible for this reduction. Also included are lysine and histidine. Hemoglobin, lactic acid, and serum albumin and globulins are also able to reduce the hexavalent chromium compounds to trivalent ones. The protein-bound chromium was always detected in its trivalent form. Only trivalent chromium compounds have a tanning and complex-forming effect. In the "two-bath" procedure of tanning, hexavalent chromium is used in the first bath; it is, however, reduced to its trivalent form by a reducing agent and hydrochloric acid of the second bath. In the "one-bath" method only trivalent chromium (chromium III sulfate) is used.

The relationship between the tanning capacity and sensitizing power of various chromium salts was experimentally investigated by Mayer and Jaconia.[149] They found that the strongest sensitizer was a complex trivalent chromium salt, which also possessed strong tanning properties. The discrepancy between the well-known strong sensitizing capacity of the hexavalent potassium dichromate and its lack of tanning properties was explained by the transformation of potassium dichromate in the body into a complex tanning salt. The other possibility which these authors suggested was, that the cause of its strong skin-sensitizing properties was its specific affinity for collagen.

The reduction of hexavalent chromium to its trivalent form with the subsequent binding to proteins may also occur intracellularly. The interaction of chromium salts with erythrocytes can be considered an example for this hypothesis. Hexavalent chromium salts permeate the membrane of red blood cells and are inside the cells reduced to a trivalent form. However, hexavalent chromium has no agglutinating effect. Trivalent chromium agglutinates erythrocytes but is unable to penetrate their membranes.[204,216a]

From the present view about the role of macrophages and Langerhans' cells in the induction and elicitation of contact sensitivity, this intracellular reduction of chromium compounds becomes even more important. It is not exactly known, whether for antigen presentation the hapten or hapten-protein conjugate have to enter the cell. However, if this is so, the chromium compounds have to penetrate into the stimulator cells in their hexavalent form since trivalent compound do not permeate the membranes. This, however, does not exclude an active phagocytosis of trivalent compounds by macrophages or Langerhans' cells.

The importance of the trivalent chromium for the sensitization was further underlined by the finding that circulating antibodies could never be traced against hexavalent but only against trivalent chromium forms.[120,194,217–220] Some authors were, however, unable to detect any antichromium antibodies either in humans or in guinea pigs.[221,222] In this context it should be mentioned that reduced chromium may sometimes be further precipitated, thus loosing its sensitizing capacity.[110,223]

It is of interest that compounds which furthered the reduction of the hexavalent chromium to its trivalent form such as carbostyril derivatives also increased the proportion of guinea pigs which became hypersensitive to potassium dichromate.[151]

There exist, however, sporadic reports in the literature, which dispute the necessity of the reduction of hexavalent to trivalent chromium and even postulate a conversion in the opposite direction, i.e., an oxidation of the trivalent to the hexavalent forms.[14,39,224] Spier and Natzel[14] suggested that this conversion of trivalent to hexavalent chromium occurs in the skin under the influence of light and oxygen. However, later on they changed their opinion on the basis of their own experiments.[174,175]

All the facts reviewed in this section strongly support the hypothesis that chromium penetrates into the skin in its hexavalent form but is reduced there to its trivalent one mainly by the sulfhydril groups of amino acids. Trivalent chromium then forms antigenic complexes by binding with proteins and this binding is, according to Landsteiner, the precondition for immunogenicity of haptens.

The exact role of stimulatory cells (macrophages of Langerhans' cells) in sensitization

Table 6
CROSS-SENSITIZATION AND CROSS-REACTIONS WITH Cr^{VI} and Cr^{III}

Sensitizing agent	Test with	% of positive animals			
		To epicutaneous test %	Mean intensity	To intradermal test (%)	Mean diameter
$K_2Cr_2^{VI}O_7$	$K_2Cr_2^{VI}O_7$	91	1.5	91	15.6
	$Cr^{III}Cl_3$	51	0.7	77	11.6
$Cr^{III}Cl_3$	$K_2Cr_2^{VI}O_7$	95	1.21	100	14.6
	$Cr^{III}Cl_3$	38	0.5	74	11.2

to chromium has been only insufficiently elucidated even if some facts indicate their importance.[169,199] However, it seems to be undisputed that the surface proteins of these cells, especially the Ia structures, play a decisive role in both the induction and the elicitation of contact sensitivity to chromium. Whether or not the determinants presented to lymphocytes consist of hexavalent or rather trivalent chromium has still to be elucidated.

D. Immunogenicity of Trivalent and Hexavalent Chromium Compounds in the Guinea Pig

In the previous section, an attempt has been made to review some chemical and physical properties of trivalent and hexavalent chromium salts with respect to their possible sensitizing capacity. Having shown substantial differences in the capacity of penetrating into the skin and of binding proteins and having stressed the possibility of conversion of hexavalent compounds to trivalent ones, it is important to correlate these differences with the sensitizing capacity in guinea pigs as the most suitable experimental model.

Groups of Hartley guinea pigs were sensitized either with 1 mg potassium dichromate or with 2 mg chromium chloride in FCA as described in the corresponding section. The amounts of elementary chromium used for sensitization were approximately the same. The animals were repeatedly challenged by the epicutaneous and intradermal way with both chromium compounds, and the intensity of the skin reactions was compared at the time when these reached their maximum, i.e., about 4 to 5 weeks after sensitization. In Table 6 it is shown that skin reactions to challenges with hexavalent potassium dichromate were always of higher intensity than reactions to trivalent chromium chloride. These results apply to both epicutaneous and intradermal challenges. However, no differences between skin reactions in animals sensitized with different compounds in relation to the compound used for sensitization could be observed provided that skin reactions were elicited with the same chromium compound. Thus guinea pigs sensitized with either hexavalent or trivalent chromium compounds gave reactions of the same intensity when challenged with hexavalent chromium. The intensity of skin reaction of these animals was lower but again the same when challenged with trivalent chromium.

This leads us to the conclusion that both hexa- and trivalent chromium salts possess an equal sensitizing capacity and that the difference between these two salts becomes apparent only during the eliciting reaction.

Most experimental data on sensitization of experimental animals to various chromium compounds are in agreement with our results and slight differences observed by some authors could be easily explained by different techniques used for either induction or elicitation of contact sensitivity.

Van Neer[159] was able to induce the same degree of contact sensitivity in guinea pigs sensitized by intradermal application of either hexavalent potassium dichromate or of trivalent chromium sulfate. Positive skin reactions were, however, elicited only by epicutaneous application of potassium dichromate. Epicutaneously applied chromium sulfate gave consistently negative results.

This may be due to a lower skin penetrating capacity of chromium sulfate in comparison to chromium chloride used in our experiments. Intradermal challenges were not used in Van Neer's experiments. Schneeberger and Forch[225] induced contact sensitivity to chromium with both tri- and hexavalent water-soluble chromium salts in 2% Triton® X-100 by repeated painting of guinea pig skin. The capacity of eliciting contact sensitivity reactions was, however, significantly higher with chromium VI than with chromium III compounds. Water-soluble chromium compounds were generally poor sensitizers and elicitors.[225]

Mali et al.[194] sensitized guinea pigs with chromium sulfate in FCA and found that this compound induced contact sensitivity of a lower degree and less frequently than hexavalent salts. However, the amount of elementary chromium used for sensitization in these experiments was about four times lower than in ours with chromium chloride. This may be the reason why in Mali's experiments the trivalent chromium compound induced contact sensitivity with more difficulties than the hexavalent one.

The skin reactions to challenges with hexavalent chromium salts applied intradermally or by iontophoresis were consistently of higher intensity than reactions to trivalent chromium.

Gross et al.[157] observed that guinea pigs sensitized with trivalent chromium salts in FCA developed contact sensitivity of the same degree as animals sensitized with hexavalent compounds. On the other hand, intradermal challenges with trivalent chromium elicited contact sensitivity reactions of a lower degree than challenges with hexavalent ones or gave even negative results. The weaker reactions to trivalent chromium can hardly be attributed to the lower penetration capacity of this compound into the skin since the application was intradermal. However, one may accept that the trivalent chromium is less easily processed by stimulatory cells such as Langerhans' cells or macrophages if processing takes place inside the cell or that the general higher binding capacity to proteins of trivalent compounds prevents them from forming the relevant antigenic complexes.

Heise et al.[147] compared, in guinea pig experiments, the sensitizing capacity of the hexavalent potassium dichromate with a complex trivalent chromium salt of the formula $[Cr_2(OH_2)SO_4]^{2+}SO_4^{2-}$. In order to enhance the sensitizing capacity of the trivalent complex salt, it was dissolved in dimethylsulfoxide. As in our experiments, both compounds sensitized equally well as was shown by the same intensity of skin reactions elicited with hexavalent chromium. Challenges with the trivalent compound did not elicit any clear and reproducible reactions.

Interesting results were reported by Shmunes et al.[226] Whereas guinea pigs sensitized with potassium dichromate responded better to challenges with this hapten than with chromium chloride, conjugates of chromium chloride with various amino acids elicited skin reactions more frequently and of higher intensity than conjugates with hexavalent chromium salts. The authors considered these results as a further evidence for the reduction of hexavalent chromium to its trivalent form before the relevant conjugates are formed.

From this brief survey it is evident that, as far as induction of contact sensitivity is concerned, trivalent chromium compounds are almost as potent sensitizers as hexavalent ones. As far as elicitation of skin reactions in contact-sensitive animals is concerned, the majority of authors yielded better results with hexavalent than with trivalent chromium, irrespective of the mode of application.

In this regard there also exist, however, contradictory results. Jansen and Berger[28] found that potassium dichromate and chromium sulfate are not only equally good sensitizers but also elicitors. Intradermal challenges with these compounds resulted in equally intense skin reactions.

Furthermore, Schwarz-Speck and Keil[73] and Schwarz-Speck[154] (who used the method of epicutaneous sensitization and challenging with chromium sulfate and potassium dichromate, both substances being dissolved in a detergent [Triton® X-100]) did not observe any differences in the degree of contact sensitivity independently on whether trivalent or hexavalent compounds were used for either elicitation or induction.

However, apart from the latter mentioned reports, the experimental results discussed above are in agreement with the majority of clinical reports that were reviewed in the corresponding section. It seems that the antigenic complexes formed by application of both hexa- and trivalent chromium salts are in the immunological respect identical, which would support the hypothesis that the conversion of hexavalent compounds to trivalent ones is of importance. The lower intensity of skin reactions to the trivalent chromium could be explained by mechanisms preventing these compounds to form relevant immunogenic complexes. This may include both a lower penetrating capacity into the skin or through membranes of stimulator cells, as well as higher binding capacity to irrelevant proteins. The available experimental and clinical data did not, however, answer the crucial question, namely, what is or are the relevant determinant(s) in chromium contact sensitivity? It might be that contact sensitivity to chromium is directed concomitantly against both tri- and hexavalent chromium determinants thus turning out to be a complex hypersensitivity to more than one antigenic structure. The other possibility is that there is a common determinant formed from both tri- and hexavalent compounds, the latter probably after conversion to trivalent.

E. In Vitro Reactions Associated with Contact Sensitivity

Since neither clinical observations nor in vivo experiments in humans and guinea pigs were able to give a final answer to the problem of the significance of the valency of chromium compounds for both the induction and elicitation of contact sensitivity, an attempt has been made to solve this problem in in vitro experiments. The advantage of these experiments is that the influence of the penetrating capacity of various compounds into the skin, as well as some of the in vivo acting regulatory mechanisms, are excluded. The penetration capacity through the cell membranes, however, plays still an important role.

During the past 15 years, three in vitro tests gained increasing importance as correlates of delayed type hypersensitivity reactions, namely the lymphocyte transformation test (LTT), the macrophage migration inhibition test (MMT), and the assay for the skin reactive factor (SRF). The assay for killer T-cells does not seem to play a significant role in contact sensitivity research.

The LTT is based on the observation that lymphocytes from sensitized individuals show an increased DNA-synthesis upon in vitro stimulation with the specific antigen. Since the activity of effector cells is not connected with the proliferating ability, the LTT seems to reveal the presence of memory cells and is also, under certain circumstances, positive in individuals not exhibiting a positive skin test reaction.[227]

The MMT, on the other hand, is based on the fact that T-effector cells release, upon in vitro stimulation with the antigen, a lymphokine called macrophage migration inhibition factor (MIF) which, under appropriate conditions, does inhibit the migration of macrophages. This test may remain positive even if lymphocyte proliferation is suppressed, e.g., by mitomycin C treatment.

The SRF is a lymphokine(s) (not precisely-defined) capable of inducing an inflammatory tuberculin-like skin reaction when injected intradermally into normal guinea pigs.

All these tests were extensively studied and proved to be reliable in vitro correlates of delayed hypersensitivity of the tuberculin type. However, as far as contact sensitivity to classic haptens is concerned, the results were erratic. This was explained by the fact that the hapten-protein conjugates formed in vivo during sensitization differed from the antigenic complexes formed in vitro in some of the mentioned tests. However, for in vitro stimulation of either memory or effector cells in contact sensitivity, the identity of both hapten and carrier is an absolute necessity.

Recently, this identity was achieved by using haptenized macrophages as stimulator cells. This method drastically improved the reliability and reproducibility of in vitro methods in contact sensitivity. Numerous papers have been published using lymph node or blood lymphocytes from guinea pigs or man successfully stimulated in vitro by dinitrophenylated or picrylated cells such as lymphocytes, erythrocytes, and especially macrophages.[229,230]

However, early in vitro experiments with metallic compounds gave only exceptionally satisfactory results. Skin explants and blood lymphocytes from normal and from nickel-hypersensitive patients did not show differing results when incubated in vitro with nickel sulfate.[231-233] On the other hand, Grosfeld et al.[234] succeeded in demonstrating blast formation of peripheral blood lymphocytes from nickel, or chromium, hypersensitive patients upon incubation with the specific hapten. In their experiments, skin explantates did not show any specific response. Blast transformation of lymphocytes from chromium-hypersensitive patients upon in vitro stimulation with potassium dichromate was furthermore demonstrated by Jung.[235]

The transformation of lymphocytes into blast cells upon incubation with metallic compounds may also occur in cells from nonsensitized individuals in a nonspecific way. Schöpf et al.[236] demonstrated that mercury compounds are capable of inducing a nonspecific in vitro transformation of normal lymphocytes comparable to the stimulation by mitogens such as phytohemagglutinin (PHA), concanavalin A (Con A) etc. This nonspecific in vitro stimulation was restricted to mercury chloride, acetate or nitrate, or to the organic mercury containing compound Merbromin. In these experiments, chromium salts or salts of other metals did not exhibit such an unspecific antigenic effect. A nonspecific in vitro transformation of normal lymphocytes by nickel acetate was reported by Pappas et al.[237] Thulin and Zachariae[228] succeeded in inhibiting migration of leukocytes from chromium hypersensitive patients with both chromium VI and chromium III conjugates to bovine serum albumin (BSA), and with chromium VI conjugates to human dermal proteins. Chromium VI-BSA proved to be the most efficient stimulator. The inhibition was antigen specific and not due to a topic effect of the antigen.[228] Recently, the presence of MIF and SRF in patients with allergy to chromium was demonstrated by Zelger and Michelmayr.[248]

Because in vitro experiments are somehow advantageous for solving important problems of chromium sensitivity, an attempt has been made to use the methods described above for solving the problem of the importance of valency for elicitation of in vitro contact sensitivity reactions. In the experiments presented in this section, our main interest was concentrated upon the eliciting capacity of chromium salts, while leaving the problem of induction to a later time. Therefore, in the majority of in vitro experiments, the guinea pigs used as a source of sensitized lymphocytes were made hypersensitive by injections of potassium dichromate in FCA as described previously.

1. Skin Reactive Factor (SRF)

The SRF was first described by Bennet and Bloom,[238] and its properties and activities were further analyzed by Pick et al.[239-241] and Maillard et al.[242] SRF is released from lymphocytes of sensitized animals upon stimulation with the specific antigen in vitro and is present in the supernatants of such cultures. An intradermal injection of 0.1 mℓ of such a supernatant induced (in normal animals) an inflammatory skin reaction corresponding to a tuberculin-type delayed hypersensitivity reaction in the sensitized animal. Injections of supernatants of cells cultured without an antigen, to which the appropriate amount of hapten was added after elimination of cells (reconstituted supernatant), or injections of a sole antigen did not induce such an inflammation. SRF is regarded as one of the mediators of delayed-type hypersensitivity albeit it is probably a mixture of different molecules with different activities.

In our recent experiments with dinitrochlorobenzene (DNCB)-contact sensitivity, it has been shown that the vascular part of the skin reaction could be elicited also by other mediators than SRF.[138] This does not, however, exclude the very important role of SRF in the elicitation of contact sensitivity inflammatory reaction under conventional conditions.

Our experiments on the production of SRF by cells from chromium hypersensitive animals were performed in the following way using essentially the methods of Bloom and Bennett[243] and Pick et al.[244]

Guinea pigs highly contact-sensitive to chromium were injected intraperitoneally with 30 mℓ sterile light paraffin oil. The animals were killed with an overdose of ether 4 days later, bled out, and peritoneal exudate cells harvested by washing out the peritoneal cavity with Earle's solution complemented with antibiotics and sodium bicarbonate (2.2 g/ℓ). After several washings, 5 mℓ of cell suspension containing 2×10^6 peritoneal exudate lymphocytes per milliliter (about 80% of peritoneal exudate cells were macrophages) was incubated in 30 mℓ Falcon® plastic tissue culture flasks at 37°C for 24 hr in a 10% CO_2 atmosphere.

The lymphocytes were stimulated by addition of 0.1 mℓ of a conjugate of chromium chloride with guinea-pig serum ($CrCl_3$-GPS) containing either 1.5, 15, or 150 μg/mℓ of the salt. Supernatants of cultures of peritoneal exudate cells incubated without addition of antigen and reconstituted with the appropriate amount of conjugate after termination of the culture, and the conjugate without cells kept under the same conditions, served as controls. After a 24 hr incubation period, the suspension was centrifuged for 15 min at 1500 g, the supernatant collected, and the control supernatant reconstituted with the antigen. The pellet was resuspended in Earle's solution, the cells disrupted by ultrasonication and centrifuged for 30 min at 25,000xg. This second supernatant containing the cell extract was also collected and the control supernatant reconstituted. SRF from lymph node lymphocytes was prepared in an analogous way. A single-cell suspension was prepared from cervical, submental, jugular, and supratrochlear lymph nodes by teasing the lymphocytes gently into Earle's solution with the help of scissors and forceps. The lymphocytes thus obtained were incubated under the same conditions as peritoneal exudate cells using, however, a higher cell density, namely, 5×10^7 lymphocytes per milliliter. The control supernatants were prepared and reconstituted in the same way as described for peritoneal exudate cells. Then 0.1 mℓ of each supernatant was injected intradermally into normal guinea pigs and the local inflammatory reaction was read after 1, 3, and 6 hr. In these experiments, the supernatant was neither

Table 7
PRODUCTION OF SKIN REACTIVE FACTOR

Supernatants of	PE cells	PE + 15γ CrCl$_3$ − GPS conjugate	15γ CrCl$_3$ − GPS conjugate	LN cells	LN + 15γ CrCl$_3$ − GPS conjugate	15γ CrCl$_3$ − GPS conjugate
Erythema	−	+	−	−	±	−
Diameter in millimeters	0	13.5	0	0	8	0
Induration in millimeters	8	9	2.5	3.5	11.5	2.5

Cell extracts of	PE cells	PE − 150γ CrCl$_3$ − GPS conjugate	150γ CrCl$_3$ − GPS conjugate
Erythema	±	++	±
Diameter in millimeters	3	10	0
Induration in millimeters	1	6.5	0

concentrated nor purified. In all experiments, the intensity and size of erythema as well as induration were assessed. The results of these experiments (3 hr reading) are presented in Table 7. The reactions were fully developed at 3 hr and almost completely vanished after 6 hr. Neither reconstituted control supernatants nor a sole antigen solution gave significantly positive results.

The results of our experiments clearly demonstrated that peritoneal exudate and blood lymphocytes from highly chromium VI hypersensitive guinea pigs are capable of releasing into culture medium an inflammatory factor(s). This factor is present in supernatants of cultures of lymphocytes stimulated with conjugates of chromium chloride with guinea pig serum as well as in extracts of in such stimulated cells. Neither reconstituted control supernatants nor reconstituted control extracts exhibited such an activity. The highest activity was present in supernatants of peritoneal exudate or lymph node lymphocytes incubated with chromium chloride-GPS conjugate containing 15 μg/mℓ chromium chloride, whereas the most active extract was obtained from cells incubated with the conjugate containing 150 μg/mℓ chromium chloride. Generally, the SRF released by peritoneal exudate cells has a slightly stronger erythema-inducing activity than SRF from lymph node cells in spite of the lower number of peritoneal exudate cells used. On the other hand, the induration induced by the SRF from lymph node lymphocytes was somehow more pronounced than the one induced by SRF from peritoneal exudate cells.

These results showing SRF activity in contact sensitivity are isolated and confirmation with other haptens is needed before final conclusions can be drawn. This concerns mainly the role of SRF as the mediator of contact sensitivity reactions, which is far from being undisputed.[138] Recently, we were able to demonstrated the SRF activity in supernatants of cultures of selected DNP-specific T-cells incubated with DNP modified macrophages.[246a]

Since (in our experiments) in vitro formed conjugates of chromium were indispensable and since these conjugates can only be formed with the trivalent but not with hexavalent chromium salts, the only conclusion as to the significance of valency which can be drawn from these experiments is that chromium III conjugates are capable of stimulating lymphocytes from chromium VI-sensitized guinea pigs. Because conjugates of hexavalent chromium could not be prepared and investigated, this model is unable to answer the question against which determinant (i.e., valency) the contact sensitivity is directed.

2. *Macrophage Migration Inhibition Test (MMT)*

In order to further approach the valency problem, both the direct and indirect macrophage migration inhibition tests, first described by David et al.[245] and Bloom and Bennet,[246] were used.

In the direct MMT, peritoneal exudate cells from chromium-hypersensitive guinea pigs were used as a source of both lymphokine producing lymphocytes and target cells (macrophages), since this exudate consists of about 10 to 20% lymphocytes, predominantly of the T-type, and 60 to 80% macrophages. The rest are granulocytes. The cells were harvested 4 days after an intraperitoneal injection of light paraffin oil by washing out the peritoneal cavity with 150 mℓ Hank's BSS. The cells were washed three times in the same solution and then a 10% suspension in minimal essential medium Eagle (MEM) was prepared. The medium was supplemented with 15% fetal calf serum and antibiotics (100 µg/mℓ streptomycin and 100 IU/mℓ penicillin). The cell suspension was drawn into microhematocrit capillaries which had been melted at one end. The capillaries were centrifuged for 5 min at 200 xg and then cut with a glass cutter precisely at the cell-liquid interphase. The cell containing part was mounted in special tissue culture chambers with the help of silicon grease. The chambers were then filled with control medium or with medium containing unconjugated haptens in appropriate concentrations.

In this direct MMT assay, sensitized lymphocytes, present in the peritoneal exudate cell suspension from hypersensitive guinea pigs, are directly stimulated by the specific hapten which probably binds serum proteins present in the culture medium. As a result of this stimulation several lymphokines, including MIF, are released. MIF immediately affects macrophages also present in the same peritoneal exudate cell suspension. MIF decreases the migration of macrophages from the capillaries in comparison to macrophages migrating in absence of this factor. For production of MIF, we used either potassium dichromate (1.5 µg/mℓ) or chromium III-chloride (150 µg/mℓ) or chromium II-chloride (300 µg/mℓ) in Eagle's MEM. These concentrations corresponded to 0.53, 29.4, and 127.8 µg of elementary chromium.

In a further series of experiments, a chromium III-chloride-GPS conjugate in a concentration of 200 µg/mℓ containing 39.2 µg elementary chromium dissolved in Eagle's MEM, was used. Peritoneal exudate cells from hypersensitive animals, cultured without hapten, and cells from guinea pigs incubated with or without hapten served as controls. Capillary-containing chambers filled with the medium containing appropriate hapten or control medium were incubated for 24 hr at 37°C and then the migration area of macrophages was assessed by projecting it on a mm-paper and counting the squares. The results were expressed as percentage of inhibition according to the formula:

$$\% \text{ inhibition of macrophage migration} = 100 - \frac{\text{migration area with hapten}}{\text{migration area without hapten}} \times 100.$$

The results of the direct MMT assay with chromium compounds of different valencies are presented in Table 8 and Figures 4 and 5.

Table 8
MACROPHAGE MIGRATION INHIBITION TEST

Antigen	% of inhibition	
	Sensitized cells	Normal cells
$K_2Cr_2O_7$ (1.5 μg)	64	4
$CrCl_3 \cdot 6 H_2O$ (150 μg)	70	7
$CrCl_2$ (300 μg)	71	11
$CrCl_3$ – GPS conjugate (200 μg)	47	3

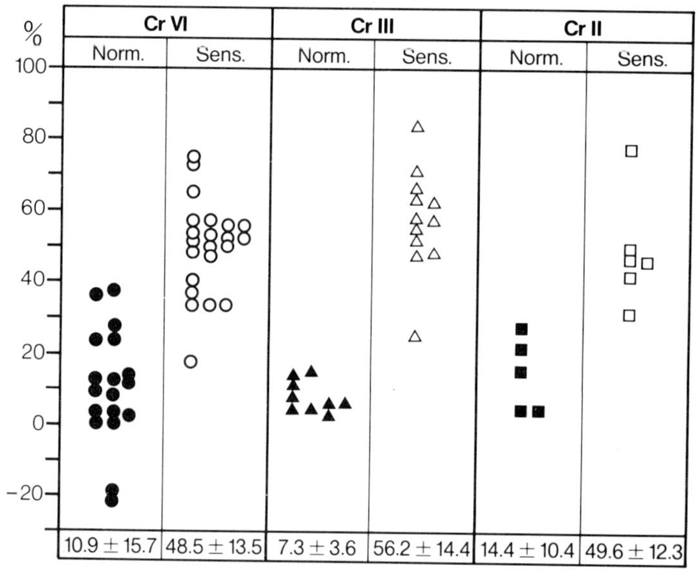

FIGURE 4. Direct MMT with chromium salts of different valencies. All guinea pigs were sensitized with potassium dichromate in FCA and the inhibitory effect induced by 1.5 μg potassium dichromate, 150 μg chromium III-chloride, or with 300 μg chromium II-chloride. The degree of inhibition was expressed as mean percent of inhibition ± standard deviation.

All chromium salts induced a significant inhibition of migration of macrophages when sensitized lymphocytes were present in the suspension. A somehow lower but still significant inhibition was also achieved by the chromium III-chloride-GPS conjugate which was also active in the previously described induction of SRF.

Peritoneal exudate lymphocytes from normal animals did not release any lymphokine upon antigenic stimulation as shown by lack of inhibition of migration of macrophages from normal peritoneal exudates. This lack of macrophage inhibition also excludes a possible toxic effect of the haptens in concentrations used.

Even if the inhibition with all three chromium salts was of about the same degree, the chromium concentrations necessary for this inhibition differed significantly. From

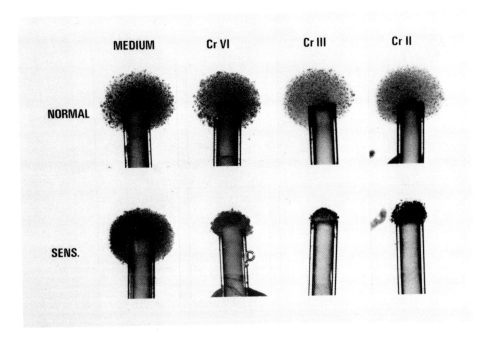

FIGURE 5. Direct MMT with chromium salts of different valencies—representative experiment.

this point of view, the required amount of the hexavalent potassium dichromate was about 100 times lower than of the trivalent chromium chloride and 200 times lower than of the divalent chromium chloride. When calculated as elementary chromium, the corresponding approximate figures were 1:50:250.

For the indirect MMT, cervical, submaxillary, axillary, femoral, subclavian, and popliteal lymph nodes from potassium dichromate contact-sensitive guinea pigs were used as source of MIF-producing lymphocytes. A single cell suspension in Eagle's MEM containing 15% guinea pig serum was prepared and 1×10^8 viable cells in 5 mℓ of this medium were incubated for 24 hr at 37°C in a humidified 10% CO_2 atmosphere in the presence of either 1.5 µg/mℓ potassium dichromate or 150 µg/mℓ chromium III chloride or 200 µg/mℓ chromium II dichloride. The cell-free supernatant obtained by centrifugation at 25,000 g for 10 min was tested for macrophage migration inhibitory effect. Peritoneal exudate cells from normal animals served as target. Supernatants from sensitized lymphocytes incubated under the same conditions but without addition of the hapten and reconstituted at the end of the culture with the corresponding concentrations of the metal salts were used as controls. MIF-activity could only be detected in supernatants from lymphocyte cultures incubated in the presence of chromium III chloride whereas supernatants of cultures in presence of potassium dichromate or chromium II dichloride did not show any inhibitory activity (Figures 6 and 7).

3. Lymphocyte Transformation Test (LTT)

The LTT gives erratic results in contact sensitivity to metals. This also applies to our unpublished attempts in which lymph node lymphocytes from chromium hypersensitive guinea pigs were cultured for 4 to 5 days in the presence of nontoxic doses of chromium III-chloride or potassium dichromate added to the culture medium. The main reason for our failure was probably the small difference between the effective and the toxic doses of chromium salts. Recently, this difficulty was at least partially overcome by

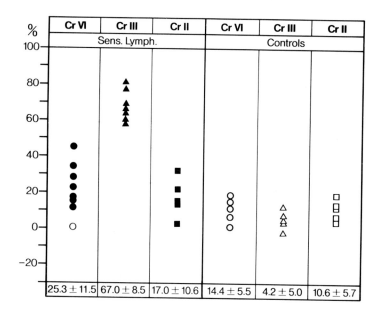

FIGURE 6. Indirect MMT with chromium salts of different valencies. Lymph node lymphocytes from potassium dichromate sensitized guinea pigs were cultured for 24 hr at 37°C with either 1.5 μg potassium dichromate, 150 μg chromium III-chloride, or 300 μg chromium II-chloride. Supernatants and control reconstituted supernatants were assayed for MIF activity. Results are expressed as in Figure 4.

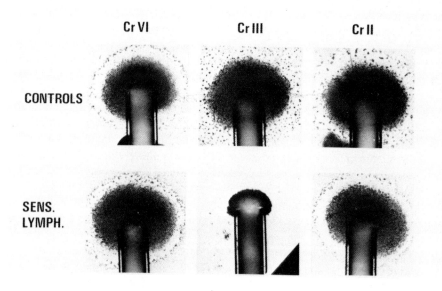

FIGURE 7. Indirect MMT with chromium salts of different valencies—representative experiment.

Table 9
INCREASED IN VITRO DNA-SYNTHESIS BY LYMPHOCYTES FROM CHROMIUM SENSITIZED GUINEA PIGS

Sensitization	$CrCl_3$/FCA		$K_2Cr_2O_7$/FCA		FCA	
In vitro stimulation	CrIII – Mϕ	CrVI – Mϕ	CrIII – Mϕ	CrVI – Mϕ	CrIII – Mϕ	CrVI – Mϕ
Restimulation						
None	N.D.		11811 (6087–20152)	3379 (1676–6784)	114 (10–217)	211 (132–290)
CrIII – Mϕ	117663 (61818–144650)	48999 (8947–106584)	41060 (34040–51173)	25321 (4680–63526)	N.D.	N.D.
CrVI – Mϕ	8589 (4588–12590)	6649 (336–12961)	8256 (5955–10556)	5807 (1054–10556)	N.D.	N.D.

Note: Lymph node lymphocytes from guinea pigs sensitized either with chromium III chloride ($CrCl_3$) or potassium dichromate ($K_2Cr_2O_7$) in FCA were assayed for in vitro DNA synthesis upon stimulation and restimulations with chromium III- or chromium VI-modified macrophages (CrIII – Mϕ; CrVI – Mϕ). Results are expressed as mean increment in cpm with minimal and maximal counts in brackets.

using chromium-modified macrophages where much lower amounts of the hapten were required to achieve an optimal effect.

As in previous in vitro experiments, guinea pigs were sensitized with 1 mg potassium dichromate in 1 mℓ FCA. Lymph node lymphocytes were harvested 14 days later and 5×10^5 cells cultured in RPNI 1640 medium supplemented with antibiotics, 1-glutamine, 2-mercaptoethanol, and 5% fetal calf serum. Chromium-modified macrophages were prepared by reacting either 3 mg/mℓ chromium chloride or 300 μg/mℓ potassium dichromate with (5 to 10×10^7/5 mℓ Hanks BSS) irradiated peritoneal exudate cells for 10 min at 37°C. After three washings 2×10^5/mℓ chromium III or VI modified macrophages were added to the culture which was pulsed 18 hr before harvest with 1 μCi ^{3}H-thymidin. All cultures were set up in tetraplicates and the results expressed as mean increment in counts per minute.

Enhanced DNA-synthesis in vitro could be demonstrated in both Cr III-Mø and Cr VI-Mø stimulated lymphocytes, the former being significantly more effective than the latter. The mean stimulation indexes were 26.3 ± 13.4 and 7.4 ± 3.9, respectively. In order to achieve in vivo a high degree of contact sensitivity, guinea pigs have to be boosted several times with intradermal and epicutaneous applications of the hapten. However, in our experiments, lymphocytes were taken only 14 days after the first sensitizing procedure and the results, described in Table 9, were not always reproducible. The sometimes negative results may be due to the insufficient degree of contact sensitivity of the guinea pigs submitted to a short sensitizing procedure.

In order to overcome this difficulty, an attempt was made to simulate the in vivo boosting by in vitro restimulations with chromium-modified macrophages. This might also help to clarify the problem whether chromium sensitivity is directed against a common chromium determinant or against differing determinants consisting of either only chromium III compound or only chromium VI compound.

For this purpose, 2×10^7 lymph node lymphocytes from either chromium chloride or potassium dichromate sensitized animals were incubated in 10 mℓ supplemented medium to which 5×10^6 of either Cr III-Mø or Cr VI-Mø were added. Twice a week, a 4 mℓ medium was exchanged and every week 5×10^6 fresh chromium modified macrophages were added. After 3 weeks the cells were harvested and triplicate cultures of 1×10^5 lymphocytes with 1×10^5 hapten modified macrophages were set up for 4 days. On day 3 the cultures were pulsed for 18 hr with 1 μCi ^{3}H-thymidin.

The results presented in Table 9 clearly demonstrate that lymphocytes from chromium hypersensitive guinea pigs are capable of enhanced in vitro DNA-synthesis providing their degree of contact sensitivity is sufficiently high. The failure of exhibiting an in vitro positive proliferative response could be attributed to the incomplete sensitization. It has already been shown in vivo that there exist substantial differences in the time required for the development of chromium contact sensitivity in guinea pigs.

In all experiments, Cr III-Mø were more efficient stimulators of lymphocytes than Cr VI-Mø. This was particularly expressed in the restimulating (booster) effect and less pronounced in the conventional induction of the enhanced DNA-synthesis in vitro. However, no significant differences were observed in respect to the chromium valency used for in in vitro sensitization. Moreover, continuous restimulations with one antigen preparation, e.g., Cr III-Mø did not abolish the capacity of the other preparation, i.e., Cr VI-Mø, to elicit an in vitro response in such treated lymphocyte suspensions. In other words, it is not possible to achieve a positive selection of cells responding only to Cr III-Mø or only to Cr VI-Mø. This is considered as strong evidence that contact sensitivity to chromium is directed against only one common determinant independently of whether hexa- or trivalent chromium compounds were used for in vivo sensitization and/or for in vitro restimulation.[255a]

F. Concluding Remarks to the Valency Problem

In order to explain the differing immunogenic capacity of chromium compounds of different valencies, which are expressed not only in vivo but also in vitro, the following properties of these compounds have to be considered:

1. Penetration capacity concerning both the skin and cell membranes
2. Binding capacity to proteins
3. Conversion of hexavalent chromium into its trivalent form under influence of skin and other proteins

Hexavalent chromium penetrates all membranes without any difficulty, while being unable to bind proteins. Trivalent chromium, on the contrary, is a strong protein binder but is unable to penetrate membranes or the skin. This has been clearly demonstrated in experiments where hexa- or trivalent chromium salts were injected intravenously. Hexavalent chromium was completely taken up by erythrocytes and other blood cells, whereas trivalent chromium was found to be entirely bound to serum proteins.

The varying sensitizing activity of chromium compounds of different valencies in vivo was therefore usually explained by their different capacity of penetrating into the skin. This might be true when the hapten is applied epicutaneously; it has a lesser value for intradermal application, but certainly cannot be made responsible for differences observed in in vitro experiments. However, since the differences in the immunogenic activity observed in vitro were of the same magnitude as in vivo, other factors must be involved. The capacity of penetrating into skin can only signify an additional furthering or limiting factor.

In view of the recent information about the role of accessory cells in the recognition of antigens and haptens by T-cells, the internalization of the haptens, i.e., their ability to penetrate cell membranes, becomes of decisive importance. It is acceptable that the relevant immunogenic complex is formed only inside the stimulatory cells and then exposed together with the Ia-antigens on the cell surface. Since the hexavalent chromium compounds penetrate the membranes more easily than the trivalent ones, much lower amounts are required for achieving the necessary concentration inside the cell. Trivalent compounds with very low penetrating capacities have to be present in the cell environment in very high amounts to reach the decisive sites inside the cells in a sufficient amount.

On the other hand, trivalent compounds are capable of binding directly to proteins whereas hexavalent ones do so probably only after conversion into trivalent. This may have two consequences:

1. Hexavalent chromium has to be reduced inside the cells to trivalent in order to be able to form immunogenic complexes. This may, however, occur without great difficulty, since all necessary sulfur containing amino acids are also present inside the cells.
2. Trivalent chromium, by virtue of its rather strong protein binding capacity, may become attached to immunologically irrelevant proteins outside the stimulatory cells, thus being prevented from reaching the immunologically important carriers inside the stimulatory cells. This may occur both in vivo and in vitro.

Another possibility was suggested by Cohen.[247] On the basis of the experiments, the author suggested that chromium compounds of all three valencies, namely hexa-, tri-, and divalent, participate in the specific immune response and that sensitivity is directed against three different determinants. In that case, the threshold values necessary for the elicitation of the immune response in vitro and in vivo may considerably differ according to the valencies. The failure to select in vitro lymphocyte populations directed against only trivalent or only hexavalent chromium compounds refutes this possibility.

An original (even if less probable) hypothesis was suggested by Everall et al.[45] According to these authors, sensitivity to chromium is directed against the elementary chromium itself and conjugation, where ever it occurs, is not a prerequisite for elicitation of the immune response. According to this hypothesis, the difference in the immunological activity of chromium compounds of different valencies is explained by different dissociation constants of various chromium compounds. This means that, at the end, for the liberation of the same amounts of elementary chromium, different amounts of chromium compounds are necessary in relation to the valencies of chromium contained in these compounds.

Interesting conclusions could be drawn from the results of the indirect MMT. When lymph node lymphocyte suspensions containing very few macrophages were used, MIF was only produced upon challenge with trivalent but not with hexa- or divalent chromium salts. This indicates that higher numbers of macrophages are necessary for processing and presentation of hexa- and divalent chromium salts than for trivalent ones. It is not clear whether this difference is due to the necessity to convert all chromium compounds to trivalent ones. If this is the case, it would mean that chromium hypersensitivity is finally directed against determinants formed with trivalent chromium. This opinion is shared by many experimental and clinical workers and is further supported by our results in the LTT. In this test, chromium III modified macrophages appeared to be the most efficient stimulators for lymphocytes from chromium contact-sensitive guinea pigs. Moreover, enhanced in vitro DNA-synthesis could be elicited in lymphocyte populations from guinea pigs sensitized either with tri- or hexavalent chromium salts. This proliferative response was obtained by stimulation with macrophages conjugated with trivalent as well as hexavalent chromium compounds. The attempt to select specific lymphocyte populations responding only to one of these antigenic preparations failed. These results clearly indicate that contact sensitivity to chromium is directed against one common determinant. This conclusion is supported by almost all available in vivo and in vitro data. However, it is difficult to reach a final decision whether this determinant consists of trivalent chromium. The results of our in vitro experiments as well as some in vivo obtained data point in this direction.[226]

The better effect of hexavalent chromium salts in vivo could be explained by the better penetration capacity of this compound. Finally, this hexavalent chromium has to be converted into trivalent in order to form immunogenic complexes. The occasionally encountered better activity of hexavalent chromium in vitro could be explained by its ability to reach (under certain conditions) the immunologically critical site more easily than the trivalent one. Besides its inferior penetrant capacity, this may be partly caused by the fact that trivalent chromium could be prevented from forming immunogenic conjugates because of binding to irrelevant proteins. This would also explain that for efficient haptenization of macrophages, higher amounts of trivalent than of hexavalent chromium are required.

VI. GENETIC FACTORS IN CONTACT SENSITIVITY TO METALS

It is a well-known clinical experience that some individuals become sensitized after a short exposure to a compound known to be a weak or medium strong sensitizer, whereas others remain resistant to a sensitization even after a long exposure to a strong immunogen. From these observations it is evident that beside the sensitizing capacity of the allergen, additional factors play an important role in the development of allergic reactions. This includes the influence of seasons, food, climate, etc.[250,264] Furthermore, the view that genetically determined factors may play a decisive role meets with steadily increasing interest. Bloch and Steiner-Wourlisch[249] succeeded in sensitizing practically all their patients with primulin, *p*-phenylenediamine, dinitrochlorobenzene, poison ivy, etc. However, when less aggressive, even though rather frequent, antigens or haptens were used to which (according to their opinion) also belongs chromium, hypersensitivity occurred in a limited number of individuals. This difference in susceptibility was at least partially determined by genetically controlled factors.[251]

In contact sensitivity, this was first experimentally proved in the work of Chase.[252] Chase set up colonies of Hartley guinea pigs by selecting the parents in each generation according to their ability to manifest contact sensitivity to dinitrochlorobenzene. After several generations of breeding, he obtained two strains of guinea pigs. One of them gave uniformly intense reactions after a brief course of sensitization (so-called Rockefeller strain). The other strain which could be sensitized but with great difficulty, developed a low degree hypersensitivity only after an intense course of sensitization.

Recently, it has been shown that T-cells, including cells which mediate contact sensitivity, are capable of recognizing an hapten (or antigen) only in presence of a genetically determined structure on the surface of antigen-presenting cells. This structure, called immune response gene associated antigen (Ia-antigen), is coded by the immune response gene (Ir-gene) which is a part of the major histocompatibility complex (MHC).[253-255] This has first been demonstrated with synthetic antigens such as dinitrophenyl conjugates of poly-L-lysine, linear copolymers of tyrosine and glutamic acid, and linear and branched copolymers containing lysine.[256-258]

Using the two inbred strains of guinea pigs, namely, strain 2 and strain 13, significant differences in their immune reactions to the mentioned antigens were shown. Furthermore, it has been shown that hapten-modified macrophages or Langerhans' cells are able to induce contact sensitivity to the specific hapten only in syngeneic Ia-antigen homologous recipients.[259]

In our laboratory, we have been studying genetic restriction of contact sensitivity to various inorganic metal compounds such as potassium dichromate, beryllium fluoride, and mercury chloride.[260] For this purpose, the outbred guinea-pig strains Hartley, Pir-

Table 10
PROPORTION OF GUINEA PIGS SENSITIZED WITH INORGANIC METAL COMPOUNDS

	Strain				
	Hartley	II	XIII	Pirbright	Himalayan
$K_2Cr_2O_7$	48/73 (66%)	8/10 (80%)	0/8 (0%)	20/30 (67%)	32/39 (83%)
BeF_2	26/33 (79%)	8/11 (72%)	0/10 (0%)	5/10 (50%)	0/10 (0%)
$HgCl_2$	7/15 (46%)	0/8 (0%)	8/10 (80%)	N.D.	N.D.

bright, and white-spotted Himalayan and the inbred strains 2 and 13 were used. We have been able to demonstrate that there exist clear differences between these strains with regard to their immune reactivity to these haptens. In the case of beryllium fluoride, the genetic dependence of these differences was experimentally proven further.

In these experiments, groups of guinea pigs of various outbred or inbred strains were sensitized to potassium dichromate with the combined method described in the previous chapters. For sensitization to beryllium fluoride, animals were painted daily on the skin of the ears with 0.05 ml of a 20% solution in Tween® 80 for 3 days. They were tested epicutaneously 14 days later with 0.05 ml of a 1% solution in 1% Triton® X-100 on the clipped flank skin. For sensitization to mercury, the left ear was painted daily epicutaneously with 0.05 ml of a 30% mercury chloride solution in ethanol for 3 days and the sensitizing procedure was repeated again after 14 days. After 14 days, the animals were epicutaneously challenged on the clipped skin of the flank with a 10% mercury chloride solution in ethanol. The solutions and solvents used for testing were tolerated without any reaction by nonsensitized animals.

The skin reactions were read 24 hr after application of the test solution and were assessed according to an arbitrary scale from 0 = no reaction to 2 = bright red and raised.

Table 10 shows that 66% of the Hartley animals became sensitized to potassium dichromate, 79% to beryllium fluoride, and 46% to mercury chloride. Of the white-spotted Himalayan animals, 83% became sensitized to potassium dichromate but none to beryllium fluoride. Of the Pirbright animals, 80% became sensitized to potassium dichromate and 50% to beryllium fluoride. With the inbred strains 2 and 13, the differences in sensitization were even more pronounced. Of the strain 2, 80% became sensitized to potassium dichromate, 72% to beryllium fluoride, but none to mercury chloride. On the other hand, 80% of strain 13 became sensitized to mercury chloride, but none to the other metal compounds.

No correlation was found in the ability of the Hartley guinea pigs to become sensitized to potassium dichromate and to beryllium fluoride (Table 11). Half of the animals reacted to both compounds in a similar way, i.e., they were either positive or negative. The other half, however, reacted to either potassium dichromate only or beryllium fluoride only.

Our experiments demonstrate that the ability to become sensitized to metal salts differs from strain to strain and that individual animals within one strain may also react to one metal compound but not to the other. This was confirmed in the Hartley and Pirbright strains and means that the reactivity to potassium dichromate and beryllium fluoride is not genetically associated. Our experiments with metal salts clearly dem-

Table 11
ABILITY OF INDIVIDUAL HARTLEY STRAIN GUINEA PIGS TO BE SENSITIZED TO $K_2Cr_2O_7$ and BeF_2

Sensitized to		Proportion of animals	%
$K_2Cr_2O_7$	BeF_2		
Pos	Pos	9/22	41
Pos	Neg	7/22	32
Neg	Pos	4/22	18
Neg	Neg	2/22	9

onstrated that there exist significant differences between various guinea-pig strains with respect to their capability of becoming contact sensitive to these haptens. The results confirm and support Chase's[252] findings that differences in sensitizability to haptens such as dinitrochlorobenzene or poison ivy also exist within an outbred strain and that they are genetically determined. However, the parallelism in sensitizability between poison ivy and dinitrochlorobenzene demonstrated in the work of Chase could not be observed between metal salts, potassium dichromate, and beryllium fluoride. It seems that genes which determine sensitization to dinitrochlorobenzene and poison ivy are more closely associated than genes for potassium dichromate and beryllium fluoride.

In order to explain why some animals within an inbred strain failed to develop contact sensitivity to a metal salt to which the majority of guinea pigs become hypersensitive, the following facts have to be considered. The guinea pigs of the inbread strain 2 and 13 are genetically homogeneous regarding histocompatibility antigens. They differ, however, in other genetically determined features such as the color of the skin and hair. Furthermore, it has been demonstrated by Bauer[207] that the rejection of skin grafts from strain 2 donors by strain 13 recipients is controlled by at least 6 genes and the rejection of strain 13 donors skin by strain 2 recipients by at least 4 genes. Recently, the structure of the major histocompatibility locus has been extensively studied in guinea pigs and the close relation between the immune response genes and the genes of the major histocompatibility locus clearly established.[253,254]

The fact that not all strain 2 guinea pigs became sensitized to potassium dichromate or beryllium fluoride and not all strain 13 animals to mercury chloride indicates that, though these strains are genetically homogeneous with respect to histocompatibility antigens, they may differ with regard to sensitization with metal compounds. Various genetically determined factors such as restricted genetic inhomogeneity, gene dose, or gene penetrance, as well as nonimmunological factors (some of them genetically determined and others not) which may influence immunogenicity indirectly, might be responsible for quantitative differences in immune response within a group of apparently syngeneic animals.

Our breeding experiments, in which the sensitizability to beryllium fluoride was used as a marker,[260] showed that in the Hartley strain, this capacity is inherited as a single Mendelian dominant characteristic. The susceptibility to become sensitized to the metal salts used in our experiments seems to be, however, controlled independently by various genes. Since Chase, in his experiments, occasionally found that some guinea pigs become sensitized either to dinitrochlorobenzene only or to poison ivy only, it might be that the immune response to these organic compounds is also controlled by different even if closely related genes.

The introduction of synthetic antigens with well-defined amino acid sequences brought about a more detailed insight into the genetic control of the cell-mediated immune response. Results of these experiments may also be helpful in understanding the genetic control of contact sensitivity to metal salts.

Ben Efraim et al.[256] observed clear-cut genetic differences in the cell-mediated reactivity of strain 2 and 13 guinea pigs to intradermal sensitization with synthetic antigens. Strain 2 animals reacted with delayed hypersensitivity to several linear and multichain polymers containing lysine, but not to a linear copolymer of tyrosin and glutamic acid. On the other hand, strain 13 guinea pigs became sensitized to tyrosine and glutamic acid polymers but not to those containing lysine. In further experiments, Ben Efraim et al.[256] demonstrated that the attachment of lysine-containing peptides to rabbit serum albumin resulted in an antigen which is reactive in strain 2 but not in strain 13 guinea pigs. These authors tried to explain the lack of immune response by a presence of clusters of positive charges within the molecule.

Levine and Benacerraf[394] tried to explain the differences in sensitizability to synthetic antigens by differences in the metabolism of these compounds. They found, however, that dinitrophenyl-polylysine conjugates are degraded in strain 13 guinea pigs, which are genetically unresponsive to this antigen, in the same way as in strain 2 guinea pigs, which are responders. Therefore they suggested that a further metabolic step, possibly specific for the lysine side-chain and controlled by a single gene, is required for the induction and elicitation of an immune response. It is possible that strain 13 guinea pigs are lacking this activating enzyme specific for the lysine side-chain which might be necessary for the coupling of the antigenic fragment to RNA.

In view of the recently described role of stimulator cells in the T-cell-mediated immune responses the enzymatic defect may alternatively relate to the ability of macrophages and/or dendritic cells to process this antigen.

The unresponsiveness of different guinea-pig strains to various sensitizers could also be explained on the basis of the "clonal selection theory" formulated by Burnet.[263] According to this theory, the inability of some animals to become sensitized to a particular antigen is caused by lack of specific antigen-reactive cells and is therefore a natural, genetically controlled, tolerance. The experimental findings in animals strongly support the assumption that the so-called individual disposition or, on the other hand, resistance so often observed in the clinic is based on a genetically controlled presence or absence of antigen-reactive cells or on the ability of stimulator cells to process and present this antigen. It may be assumed that the well-known and experimentally documented disposition or resistance to chromium is caused by similar mechanisms, i.e., by insufficient processing and presentation of the hapten to the specific T-cells by stimulator cells or by lack of these hapten-reactive specific lymphocytes. However, this assumption has not yet been experimentally proven for metal compounds.

VII. DESENSITIZATION AND TOLERANCE IN CHROMIUM CONTACT SENSITIVITY

Chromium salts exhibit, together with other heavy metals such as arsenic, some peculiarities not encountered with other nonmetallic haptens.

Contact sensitivity is, according to our present knowledge, defined as an allergic (immunologic) phenomenon mediated by a specific subpopulation of T-cells (T-effector cells) and their products. Antibodies do not seem to play any important role in this phenomenon.[122]

Attempts to prevent the development of delayed or contact sensitivity in experimental animals go back to the early experiments of Sulzberger,[265] who succeeded in inducing specific immunological unresponsiveness to neosalvarsan by injecting this arsenic-containing compound intravenously to normal guinea pigs. A subsequent sensitizing attempt in so treated animals failed to induce delayed (tuberculin-type) hypersensitivity. About 20 years later, Chase[266] induced specific unresponsiveness to picryl chloride or dinitrochlorobenzene by feeding the respective hapten to guinea pigs a certain time before the sensitizing procedure. This phenomenon of specific immunological unresponsiveness, which became known as immunological tolerance through the work of Medawar,[267,268] was then further analyzed in the contact sensitivity system by Frey,[269] de Weck,[270] Polak,[271] Asherson,[272] Claman,[273] and many others, and recently reviewed by Polak.[122] According to our present knowledge, tolerance is based either on clonal deletion, i.e., destruction and/or absence of the specific T-cell subpopulation or on a suppressive mechanism consisting of one or more subpopulations of specific suppressor cells and their products called suppressor factors.

Generally it has been demonstrated that tolerance to contact sensitivity is induced in nonsensitized (naive) individuals (guinea pigs, mice) with a relative ease by feeding or intravenously injecting the respective hapten. This tolerance is (at least in guinea pigs) complete, permanent, and long-lasting and mediated by suppressor cells. On the other hand, a loss of the immune response to an antigen or hapten to which the individual has been already sensitized is still only exceptionally achieved. This phenomenon is called, in contrast to tolerance, desensitization.

A. Desensitization

Procedures that were successful in inducing permanent tolerance in nonsensitized animals failed to induce unresponsiveness in already sensitized guinea pigs or induced an only short-lasting inhibition. Therefore, it has been concluded that different mechanisms may be operating in tolerance and desensitization.[274]

Many authors have shown that the mechanism of desensitization is, dependent to a large extent, on the antigen or hapten used and may involve macrophage inactivation,[275] humoral suppressor factors,[276–278] depletion of specifically reactive lymphocytes,[279] or on the contrary, their enhanced activation[280] and activation of suppressor cells.[281,282] Furthermore, it has been shown that the temporary desensitization of guinea pigs to dinitrochlorobenzene by an i.v. injection of this hapten was due to a transient blockade of the specific receptor on T-effector cells.[283] The reappearance of contact sensitivity was mediated by generation of effector cells newly formed from memory cells present in the lymph nodes. These latter cells are largely resistant to the inactivating effect of the intravenously injected hapten. Feeding the hapten failed completely in inducing even a slight degree of desensitization.

Because of the potentially extreme importance of the desensitization phenomenon for the future therapy of allergic contact eczema which, in spite of immense efforts, still remains unsolved, a short survey of desensitizing procedures may be of use. Specific but also short-lasting desensitization was achieved in the guinea pig and man by repeated applications of the specific hapten.[284–286]

Dinitrochlorobenzene contact-sensitive guinea pigs became specifically unresponsive when treated daily epicutaneously with this hapten for 40 to 100 days. This desensitization lasted for several weeks but was reversed by a second sensitizing attempt. The suggested mechanism for this type of unresponsiveness was a temporary exhaustion of sensitized effector cells.[287]

An i.v. injection of the highest tolerated dose of dinitrochlorobenzene or its derivative, dinitrobenzenesulfonic acid sodium salt, induced a partial or complete desensitization to this hapten. Contact sensitivity recovered, however, within 1 week but was, in comparison to the initial degree, slightly but statistically significantly diminished.[288] This difference was due to the additional effect of suppressor cells activated by the intravenously injected hapten.[139]

A specific and permanent desensitization to a hapten was reported first by Sulzberger[265] in 1929 who used neosalvarsan. His results were then confirmed and further elaborated by Frey and co-workers.[290] They succeeded in inducing a prolonged and even permanent state of specific immunologic unresponsiveness in guinea pigs exhibiting delayed hypersensitivity (tuberculin-type) to neosalvarsan. The desensitizing procedure consisted of an i.v. injection of a large dose of the hapten, followed by an intradermal injection of a smaller dose of the same substance 6 hr prior to or within 48 hr after the i.v. injection. A smaller i.v. dose or a longer time interval between the intravenous and intradermal applications, or omission of the intradermal application, converted the permanent desensitization into a temporary one, similar to the one achieved in the DNCB-system.

1. Desensitization to Chromium Compounds

As far as contact sensitivity to chromium salts is concerned, several authors attempted to achieve a permanent desensitization of patients with cement eczema by repeated epicutaneous application of potassium dichromate over a period of up to 5 months.[42,43,291-293] These experiments met with partial success. Feeding small doses of potassium dichromate to patients with chromium contact sensitivity led to an exacerbation of old inflammatory sites rather than to a desensitization.[80,81] Reports about a successful desensitization of patients with chromium eczema by this method could never be reproduced and moreover are toxicologically harmless.[294] Since Chase's[295] experiments feeding the hapten (DNCB) induced tolerance in naive animals but failed to induce an even partial or transient desensitization in already sensitized guinea pigs, the failure of this method in man does not surprise.

In preliminary studies it has been shown that an i.v. injection of large doses of potassium dichromate in guinea pigs, which have been already sensitized to this hapten, induced a prolonged state of specific unresponsiveness similar to that described by Sulzberger[265] and Frey et al.[290] for neosalvarsan. It was therefore decided to further study the effect of the i.v. injection of different doses of the chromium salt and also to find out to what extent the unresponsiveness could be modified by prolonging the time between the i.v. injection of the compound and the subsequent application of potassium dichromate to the skin.

Hartley guinea pigs were sensitized to potassium dichromate, as described previously so as to show a high degree of contact sensitivity. Thereafter they were injected intravenously through the marginal vein of the ear with potassium dichromate in doses of 20, 10, 2, 0.1, or 0.01 mg/kg dissolved in 1 or 2 mℓ of 0.15 M sodium chloride. The i.v. injection into potassium dichromate contact-sensitive animals was followed by a first skin challenge with 25 µℓ of a 0.5% solution of potassium dichromate in 1% Triton® X-100.

This challenge was applied either 6, 24, or 48 hr; 4 days; 2 weeks; or 3 months after the i.v. injection. The epicutaneous applications were then repeated 48 and 96 hr and 1 week later and then weekly until the end of the experiment. In an additional experiment, 120 mg/kg of the trivalent chromium sulfate was injected intravenously into guinea pigs sensitized to potassium dichromate. This dose contained more than twice as much elementary chromium as the highest dose of potassium dichromate (20 mg/kg).

The results in potassium dichromate contact-sensitive guinea pigs were also compared with results in guinea pigs sensitized to beryllium fluoride. In these experiments, highly contact-sensitive guinea pigs were injected intravenously with 5 mg/kg beryllium lactate dissolved in 2 mℓ of 0.15 M sodium chloride. The animals were challenged epicutaneously 24 hr later with an 0.5% solution of beryllium fluoride in 1% Triton® X-

Table 12
UNRESPONSIVENESS OF ANIMALS PREVIOUSLY SENSITIZED TO POTASSIUM DICHROMATE, FOLLOWING INTRAVENOUS INJECTION OF 20 mg/kg POTASSIUM DICHROMATE

Interval between i.v. injection and first epicutaneous skin test	Proportion of animals responsive when tested after											
	Hr			Days			Months					
	6	24	48	4	7	14	1	2	3	4	5	6
6 hr	0/6	0/6	0/6	0/6	0/6	0/6	0/6	0/6	0/5	0/5	0/4	0/4
24 hr		0/4	0/4	0/4	0/4	0/4	0/4	0/4	0/4	0/4	0/4	0/4
48 hr			0/10		5/9	5/9	6/9	5/9	6/9	5/9	6/8	6/8
4 days				0/9	2/8	4/8	4/8	5/8	7/8	7/8	7/7	7/7
14 days						3/10	7/10	6/9	9/9	7/7	7/7	7/7
3 months									7/9	7/9	6/8	3/3
Control (no i.v. injection)							8/8	7/7	7/7			

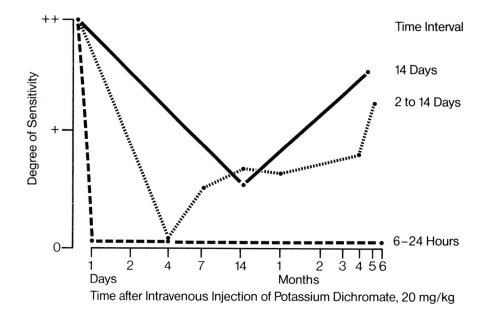

FIGURE 8. Desensitization of potassium dichromate hypersensitive guinea pigs by an i.v. injection of 20 mg/kg of this hapten. Dependence on the time interval between the i.v. injection and the first epicutaneous challenge.

100 and these challenges were repeated in the same way as was the case with potassium dichromate.

In the first set of experiments, the dependence of the desensitization on the time elapsed between the i.v. injection and the first skin challenge with the hapten was investigated and the results presented in Table 12 and in Figure 8.

In these experiments, highly potassium dichromate contact-sensitive guinea pigs were injected intravenously with 20 mg/kg of this hapten. It should be mentioned that this dose was at the level of LD_{50} and therefore the highest possible.

When the skin challenge by an epicutaneous application of this hapten was performed

Table 13
UNRESPONSIVENESS OF ANIMALS PREVIOUSLY SENSITIZED TO POTASSIUM DICHROMATE FOLLOWING INTRAVENOUS INJECTION OF DIFFERENT DOSES (INTERVAL BETWEEN INTRAVENOUS INJECTION AND FIRST EPICUTANEOUS TEST UP TO 24 hr)

Dose of potassium dichromate injected (mg/kg)	Proportion of animals responsive when tested after				
	24 hr	1 week	2 weeks	4 weeks	6 months
20	0/10	0/10	0/10	0/10	0/8
10	0/6	5/6	—	6/6	—
2	4/7	6/7	7/7	—	—
0.1	4/7	7/7	—	—	—
0.01	7/7	—	—	—	—

within the first 24 hr after the desensitizing injection, none of the animals responded to this challenge and also failed to respond to the following challenges performed at the indicated time intervals. Thus, a complete desensitization lasting over the whole time of observation, i.e., up to 6 months was achieved in all guinea pigs.

With the increasing time interval between the i.v. injection of the hapten and its first epicutaneous application, the number of guinea pigs regaining contact sensitivity to potassium dichromate during the observation time also increased even if they did not yet respond to the first challenge but only to some of the following ones. None of the animals sensitized to potassium dichromate and injected intravenously with the same hapten was permanently desensitized when the first epicutaneous application was delayed over 48 hr after the i.v. injection. When the first challenge was delayed over 2 weeks some of the animals started to react even to this first one. Only 2 out of 9 animals failed to respond to the first challenge performed 3 months after the i.v. injection, all the others being not at all desensitized.

As far as the intensity of the reactions is concerned, the initial level of contact sensitivity was, however, never reached even if the first challenge was delayed to 14 days after the i.v. injection.

The unresponsiveness thus achieved was always immunologically specific therein that guinea pigs unresponsive to potassium dichromate retained their ability to react to another hapten or antigen to which they have been previously sensitized, e.g., oxazolon or tuberculin.

The duration of desensitization was further dependent on the dose of intravenously injected potassium dichromate. This was demonstrated in the second set of experiments (Table 13, Figure 9). Keeping the time interval between the i.v. injection of the hapten and its epicutaneous application constant at the optimal time, i.e., 6 to 24 hr, the desensitizing effect was found to be markedly dependent on the dose of the intravenously injected potassium dichromate. Whereas the unresponsiveness induced with a dose of 20 mg/kg (LD_{50}) persisted for at least 6 months in all guinea pigs, most of the animals regained contact sensitivity to this hapten within 1 week when injected with half of this dose, i.e., 10 mg/kg, although all of them were initially unresponsive (24 hr after the i.v. injection). An i.v. injection of 2 mg/kg induced an only transient desensitization lasting 1 or 2 weeks in about half of the animals, the others remaining contact-sensitive. An i.v. injection of 0.1 mg/kg induced a short-lasting unresponsiveness in three out of seven animals injected. All guinea pigs responded to a second challenge performed 48 hr after the desensitizing procedure. A dose of 0.01 mg/kg failed to affect the degree

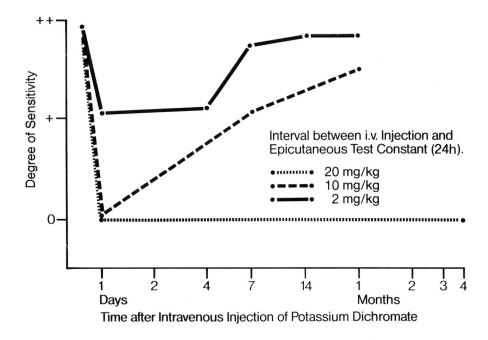

FIGURE 9. Desensitization of potassium dichromate hypersensitive guinea pigs by an i.v. injection of this hapten. Dependence on the intravenously injected dose. Interval between the i.v. injection and the first epicutaneous challenge kept constant at 24 hr.

of contact sensitivity in all of the seven treated animals.

An i.v. injection of 120 mg/kg of trivalent chromium sulfate into potassium dichromate-sensitized animals caused only a short-lasting desensitization. The unresponsiveness lasted 24 to 48 hr and a week later, all animals reacted to a skin challenge with potassium dichromate. Similar results were obtained in the beryllium system. Following an i.v. injection of 5 mg/kg beryllium lactate into 14 beryllium-sensitive guinea pigs, all animals were found unresponsive to the first skin challenge 24 hr later, but 8 out of 14 had become responsive again when skin-challenged 48 hr later, and all surviving guinea pigs reacted to an epicutaneous challenge with beryllium fluoride performed 96 hr after the i.v. injection. The dose of 5 mg/kg beryllium lactate killed 30% of the animals and was therefore the highest dose which could be used in these experiments.

The conclusion of the described experiments is that a high i.v. dose of potassium dichromate, which is near to the LD_{50} (20 mg/kg), combined with an epicutaneous application of this hapten caused a permanent desensitization in already sensitized guinea pigs and that this desensitization is dependent on the dose of the intravenously applied hapten and on the time elapsed between the i.v. injection and the epicutaneous application of the hapten. Our results confirmed the so-called "double-shot" phenomenon which was described and analyzed in the neosalvarsan system by Frey et al.[290] On the other hand, the results obtained with chromium sulfate injected intravenously into potassium dichromate contact-sensitive guinea pigs and with beryllium lactate injected into beryllium fluoride contact-sensitive animals form a parallel to the experiments in the dinitrochlorobenzene system.[269]

As it has been found that an epicutaneous challenge within 24 hr after the i.v. injection was indispensable for induction of a prolonged or permanent state of unresponsiveness, an attempt was made to analyze the action of this application.

Guinea pigs exhibiting a high degree of contact sensitivity to potassium dichromate

Table 14
UNRESPONSIVENESS OF ANIMALS PREVIOUSLY SENSITIZED TO
POTASSIUM DICHROMATE, FOLLOWING AN INTRAVENOUS
INJECTION OF 20 mg/kg POTASSIUM DICHROMATE, FOLLOWED BY
AN EPICUTANEOUS TEST WITHIN 24 hr, IF THE EPICUTANEOUS TEST
SITE IS REMOVED 6, 12, and 24 hr AFTER TESTING

Time of removal of skin test site (hr)	Proportion of animals showing skin reactivity after				
	Days		Months		
	1	14	1	2	3
6	0/6	4/6	4/6	5/6	5/6
12	0/6	4/6	4/6	5/6	5/6
24	0/6	1/6	2/6	2/6	2/6

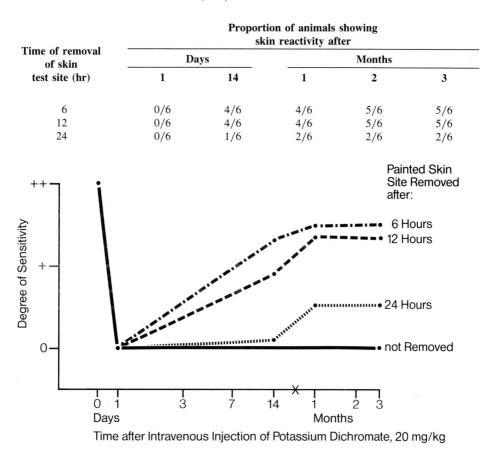

FIGURE 10. Desensitization of potassium dichromate hypersensitive guinea pigs by an i.v. injection of 20 mg/kg of this hapten. Effect of the removal of the epicutaneous hapten application site at different times after the challenge.

were injected intravenously with 20 mg/kg of this hapten, followed within 24 hr by an epicutaneous application of a 0.5% potassium dichromate solution in 1% Triton® X-100. This treatment would normally induce a complete unresponsiveness for at least 6 months. In this experiment, however, the site of hapten application was removed at different time intervals after the application of the contact agent to the skin. Removal of the skin site where the hapten was applied, at 6 and 12 hr after its application, prevented the development of a permanent desensitization. The animals began under these conditions to react within 14 days after the i.v. injection and became progressively more sensitive during the ensuing 3 months. If the skin challenge site was, however, left intact 24 hr, only two out of six animals showed skin reactivity during the next 3 months and this reaction was very weak (Table 14, Figure 10). The rest of the guinea pigs remained permanently desensitized.

The mechanism of the function of the hapten in the skin for induction of a permanent desensitization is not yet clear and will be subject to further discussion.

2. Effect of Immunosuppressive Agents

In the experiments described above, it was found that a permanent state of unresponsiveness in guinea pigs contact-sensitive to chromium could only be produced with a dose as high as 20 mg/kg potassium dichromate. This dose was, however, in the range of the LD_{50} and therefore of no practical value as a potential therapy of chromium sensitivity. Smaller doses produced a temporary desensitization in only a proportion of the guinea pigs.

Therefore, it has been considered useful to investigate whether it was possible to convert the temporary desensitization produced by an i.v. injection of lower doses of potassium dichromate into a permanent state of unresponsiveness. It may be assumed that the temporary state of unresponsiveness induced by lower doses of the intravenously injected hapten is due to an inactivation (receptor blockade) of circulating sensitized effector lymphocytes as is the case in the DNCB system.[139] The ability of the specific precursors and/or memory cells to regenerate new effector lymphocytes was not impaired by this procedure. It was possible that under these conditions, an additional application of immunosuppressive agents could lead to a selective destruction of these memory cells, which were brought to proliferation by an antigenic stimulus, e.g., by the intravenously injected hapten. Such a combined treatment may well induce a prolonged state of specific desensitization comparable to that achieved by a high dose of the hapten alone.

From the work of Turk and Stone[133] it is known that three immunosuppressive agents, in particular cyclophosphamide (CY), methotrexate (MTX), and heterologous anti-lymphocyte (ALS) or antithymocyte serum (AThS), exhibit a marked immunosuppressive effect in guinea pigs. These agents, when applied in an appropriate dose and at a suitable time, were able to completely block the capacity of specifically triggered lymphocytes to proliferate in the lymphoid organs.[133,297-300] The nonspecific effect of these agents is, by means of a simultaneous specific stimulation exerted by the hapten, converted into a specific one.

In our experiments, an attempt has been made to induce a permanent desensitization to potassium dichromate by intravenously injecting 2 mg/kg of this hapten, which itself induces a transient and incomplete state of unresponsiveness, together with the mentioned immunosuppressive agents. The agents were used either individually or in combination. The doses used in our experiments were 10 mg/kg MTX, 20 mg/kg CY, and 2 mℓ/kg ALS. All these immunosuppressive agents were administered intraperitoneally daily for 6 days, starting 2 days before the i.v. injection of 2 mg/kg potassium dichromate.

Heterologous ALS was prepared in New Zealand white rabbits by two i.v. injections of 10^9 washed fresh guinea-pig thymus cells with an interval of 2 weeks. The rabbits were bled out 1 week after the last injection and the serum absorbed with an equal volume of homologous erythrocytes. The ALS was sterilized by a millipore filtration.

Despite the fact that the guinea-pig lymphocytes are rather resistant to corticosteroids,[301] prednisolone phosphate (4 mg/kg) was used in some experiments in addition to the other immunosuppressants in an attempt to enhance the desensitizing effect and prolong the time of unresponsiveness. In all experiments, the direct effect of immunosuppressive agents on the development of the specific skin reaction to potassium dichromate in hypersensitive animals was assessed as a control. The results of these experiments are expressed as the proportion of positively reacting guinea pigs over total

Table 15
UNRESPONSIVENESS OF ANIMALS PREVIOUSLY SENSITIZED TO POTASSIUM DICHROMATE, FOLLOWING AN INTRAVENOUS INJECTION OF 2 mg/kg POTASSIUM DICHROMATE, FOLLOWED BY AN EPICUTANEOUS TEST WITHIN 24 hr, IN GUINEA PIGS TREATED WITH IMMUNOSUPPRESSIVE DRUGS

Drugs	Proportion of animals showing skin reactivity after						
	Days				Months		
	1	3	7	14	1	2	3
Methotrexate	0/9	5/9	7/8	4/5	5/5		
Cyclophosphamide	0/9	0/9	7/9	8/8	8/8		
ALS	3/5	5/5	5/5				
Cyclophosphamide + ALS	1/10	1/10	1/8	1/8	3/8	3/8	8/8
Cyclophosphamide + ALS + prednisolone	0/13	0/13	4/11	4/11	4/7	3/3	

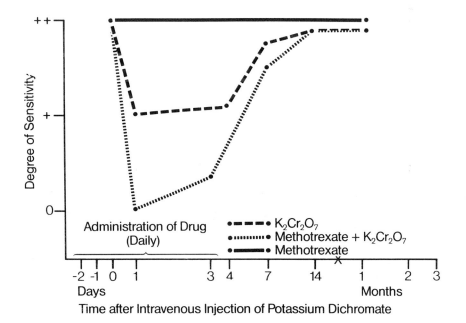

FIGURE 11. Desensitization of potassium dichromate hypersensitive guinea pigs by an i.v. injection of 2 mg/kg potassium dichromate combined with MXT.

(Table 15) and as the mean degree of contact sensitivity (Figures 11 to 15) at various times after desensitization.

a. The Effect of Methotrexate (Table 15, Figure 11)

A course of MTX, starting 3 days before the epicutaneous application of the hapten, did not influence the degree of contact sensitivity in sensitized animals challenged epicutaneously with potassium dichromate. In sensitized guinea pigs, injected intrave-

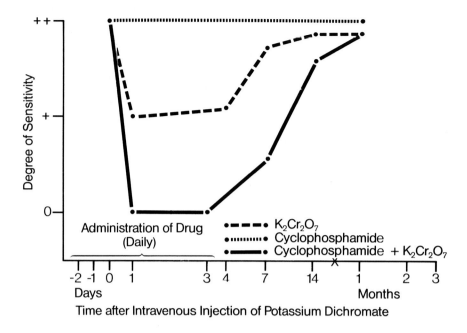

FIGURE 12. Desensitization of potassium dichromate hypersensitive guinea pigs by an i.v. injection of 2 mg/kg of this hapten combined with CY.

nously with 2 mg/kg potassium dichromate, the course of MTX (10 mg/kg) starting 2 days before the i.v. application increased the degree of suppression of the skin reactions while animals were on the drug, but did not prolong the time of unresponsiveness.

Since the dose of MTX used was of the order of LD_{50}, for this drug it appeared to be difficult to use it in a combination with another immunosuppressant and therefore further experiments with this compound were discontinued.

b. The Effect of Cyclophosphamide (Table 15, Figure 12)

The CY course, starting 3 days before the skin challenge, did not decrease the ability of the animals to manifest contact sensitivity reactions to potassium dichromate. The transient desensitization induced by i.v. injection of 2 mg/kg of potassium dichromate was increased by a CY treatment and slightly prolonged. This prolongation of desensitization persisted for about 1 week after cessation of the drug treatment. The dose of CY used in this experiment was in order of LD_{10}.

c. The Effect of Antilymphocyte Serum (Table 15, Figure 13)

Injection of ALS, starting 3 days before a skin challenge with potassium dichromate, neither suppressed the contact sensitivity reaction to this hapten nor had an effect on the degree of unresponsiveness produced by the i.v. injection of 2 mg/kg potassium dichromate.

d. The Effect of a Combination of Cyclophosphamide and Antilymphocyte Serum (Table 15, Figure 14)

When both CY and ALS were simultaneously given each day, starting 3 days before the skin challenge, to potassium dichromate hypersensitive guinea pigs, a slight and short-lasting diminution in the intensity of the skin reactions could be observed. The degree of unresponsiveness induced in such animals by an i.v. injection of 2 mg/kg

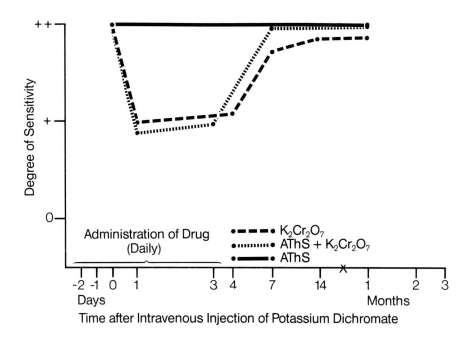

FIGURE 13. Desensitization of potassium dichromate hypersensitive guinea pigs by an i.v. injection of 2 mg/kg potassium dichromate combined with AThS.

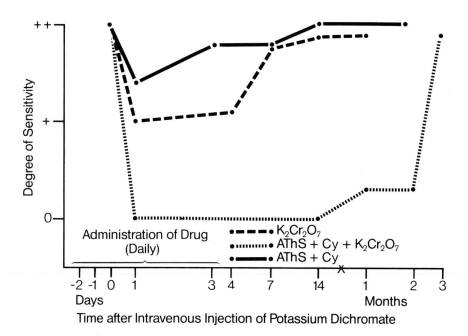

FIGURE 14. Desensitization of potassium dichromate hypersensitive guinea pigs by an i.v. injection of 2 mg of this hapten combined with CY and AThs.

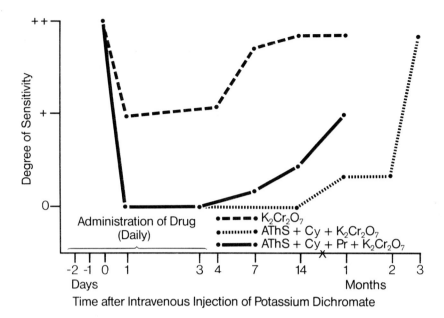

FIGURE 15. Desensitization of potassium dichromate hypersensitive guinea pigs by an i.v. injection of 2 mg/kg potassium dichromate combined with AThS and CY. Effect of an additional application of prednisolone.

potassium dichromate was, however, significantly increased and prolonged by this combination. The desensitization persisted for 2 months after the i.v. injection but there was a sharp return to responsiveness 1 month later.

e. The Effect of Addition of Prednisolone to the Combination of Cyclophosphamide and Antilymphocyte Serum (Table 15, Figure 15)

Prednisolone was added to the previous combination in an attempt to further increase and prolong the effect of both these immunosuppressive agents. However, the effect of adding this third compound to the regime was not prolongation but, on the contrary, reduction of the duration of the unresponsiveness obtained with CY and ALS.

This effect is interesting and analogous to the results described by Simmons et al.[302] They found that cortisone, when administered before or during the application of ALS in mice, decreased the degree of immunosuppression induced by ALS. In the opinion of these authors, cortisone blocks the mitogenic effect of ALS. It is, however, also possible that prednisolone inhibits the cytotoxic effect of ALS on sensitized lymphocytes.

The results obtained with a combination of CY and ALS clearly demonstrate that some immunosuppressive drugs, when given at the time of the desensitizing procedure, are capable of enhancing and prolonging the desensitizing effect. When given individually, these immunosuppressive drugs had no or only an insignificant effect on the duration of the desensitization induced by a low i.v. dose of the hapten. The effect of both these immunosuppressive drugs therefore, seems to be a potentiating one rather than just an additive one. The effect of the combination was, however, paradoxically decreased by addition of prednisolone.

B. Induction of Immunological Tolerance to Chromium Compounds

Tolerance to contact sensitivity was successfully induced by an i.v. injection or feeding of the respective hapten to naive animals at various times prior to the sensitizing procedure. An outline of these methods was summarized for guinea pigs by de Weck et al.,[270] Polak and Frey,[303] and Polak[122] and for mice by Asherson[272] and by Claman.[273] A complete and permanent tolerance to DNCB-contact sensitivity was also intravenously induced by a simultaneous application of the hapten for tolerance and intradermally in FCA for sensitization. Tolerance was still induced when the hapten was injected intravenously not later than 24 hr after the sensitizing attempt. Since the tolerogenic application was performed at the time of sensitization, i.e., during the primary response, the resulting specific unresponsiveness was called tolerance during the primary response.[304,305] In further experiments, it has been shown that the mechanism of both types of tolerance, i.e., tolerance by pretreatment and tolerance during the primary response, is practically identical and mediated by suppressor cells.[308]

Tolerance to chromium contact sensitivity was also induced by intradermally injecting potassium dichromate homogenized in Freund's incomplete adjuvant.[306] Because the authors did not further analyze this phenomenon, the mechanism remains obscure. It may be related to the tolerance induced by pretreatment with sole Freund's incomplete adjuvant as reported by Jankovic.[307]

In preliminary experiments, we were able to confirm the tolerogenic effect of the intravenously injected hapten (potassium dichromate) for contact sensitivity to chromium. Guinea pigs injected intradermally with 10 mg/kg potassium dichromate on two occasions, 28 and 48 days prior to the sensitizing attempt with this hapten, did not show any traces of skin reactions to repeated intradermal or epicutanous challenges. This tolerance was complete and lasted for the whole time of observation, i.e., over 3 months.

In further experiments, we investigated the dose of intravenously injected potassium dichromate which was required for induction of tolerance during the primary response. Since 20 mg/kg of this hapten was capable of inducing a permanent state of unresponsiveness even in already sensitized animals, only lower doses were investigated. Guinea pigs of the Hartley strain weighing between 400 to 500 g were intravenously injected with either 2, 5, or 10 mg/kg potassium dichromate and at the same time sensitized with 1 mg/kg of this hapten in FCA as described previously. The guinea pigs were repeatedly challenged once a week and the final reactions read at the time when all controls were strongly positive, i.e., 28 days after the i.v. injection for intradermal challenges and 49 days for epicutaneous challenges. From the results of these experiments, expressed in Table 16 and Figure 16, it is evident that for induction of tolerance during the primary response, only half of the dose which was necessary for induction of a permanent desensitization (10 mg/kg) was required to achieve a long-lasting tolerance in about 80% of the animals. However, even 5 mg/kg of potassium dichromate injected intravenously still induced a state of tolerance in about 50% of the guinea pigs treated. The tolerant animals were equally unresponsive to epicutaneous as to intradermal challenges.

The view may be held that in hypersensitive individuals the number of antigen-reactive effector lymphocytes is largely increased. From this point of view it is understandable why, for induction of permanent unresponsiveness in naive animals, lower amounts of haptens are required than for induction of desensitization in already sensitized guinea pigs. It seems that quantitative relations exist between the number of effector cells to be blocked by the desensitizing procedure and the dose of the hapten applied intravenously for this inactivation (blockade). This is supported by the findings of Frey et al.,[304] who induced a state of tolerance during the primary response to neosalvarsan in the majority of naive guinea pigs with one tenth of the dose necessary for inducing a permanent desensitization (7.5 mg/kg neosalvarsan).

The different doses of hapten required for induction of tolerance and for induction of desensitization may, however, be even more satisfactorily explained by the differ-

Table 16
INDUCTION OF IMMUNOLOGICAL TOLERANCE DURING THE PRIMARY RESPONSE TO POTASSIUM DICHROMATE

			Intradermal test		Epicutaneous test	
			$K_2Cr_2O_7$ 25 µg		$K_2Cr_2O_7$ 100 µg	
Experiment	Sensitization 1 mg $K_2Cr_2O_7$ Intramuscularly Day 0	Tolerization $K_2Cr_2O_7$ intravenously mg/kg	Animals sensitized/ total	Mean diameter of reactions (mm) 28	Animals sensitized/ total	Mean degree of reactions 49
a	1 mg	10	1/7	10.00	2/7	0.35
b	1 mg	5	5/9	10.65	5/8	0.75
c	1 mg	2	8/12	13.25	11/12	1.65
d	·/·	10	0/4	7.0	0/4	0
e	·/·	5	0/3	7.0	0/3	0
f	·/·	2	0/4	8.7	0/4	1.6
g	1 mg	·/·	25/32	15.3	25/30	0
h	·/·	·/·	0/8	6.5	0/8	

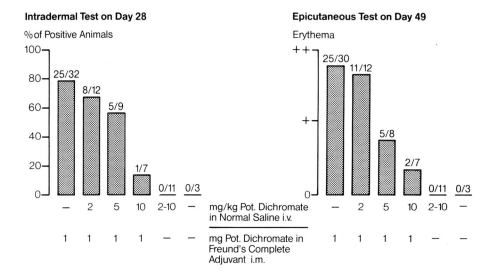

FIGURE 16. Induction of tolerance during the primary response in guinea pigs hypersensitive to potassium dichromate. Dependence on the intravenously injected dose of hapten.

ences in the mechanisms mediating these two phenomena.

In the DNCB-system in guinea pigs, it has been shown that tolerance induced during the primary response and tolerance by pretreatment are mediated by suppressor cells,[179,304,305] whereas desensitization is based on a long-lasting blockade of the receptors of effector cells.[283] In the state of permanent desensitization, as is the case in chromium sensitivity, an additional destruction or permanent inactivation of existing memory cells and suppression of their regeneration must also be involved.

It may be postulated that for activation of suppressor cells (induction of tolerance by pretreatment), lower amounts of the tolerogen are required than for the receptor blockade or inactivation of memory cells (desensitization). These are, however, only indirect hypotheses, since the actual mechanism of tolerance and desensitization in chromium

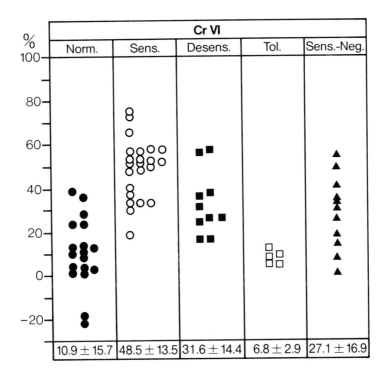

FIGURE 17. Direct MMT with peritoneal exudate cells from animals sensitized (sens), desensitized (desens), tolerant (tol), and from animals submitted to sensitizing procedure but not exhibiting positive skin reactions (sens-neg).

contact sensitivity has not yet been experimentally assessed.

The unresponsiveness of lymphocytes from tolerant or desensitized guinea pigs was further investigated in both the direct and the indirect MMT. We were able to demonstrate that peritoneal exudate cells from tolerant guinea pigs behaved in the direct MMT like cells from normal nonsensitized guinea pigs, i.e., their migration was not inhibited in the presence of the specific hapten (Figures 17 and 18). Moreover, supernatants from cultures of lymph node lymphocytes from tolerant guinea pigs incubated with the specific hapten did not exhibit MIF activity, i.e., did not inhibit the migration of macrophages from capillaries (Figure 19). A somehow unexpected result emerged from experiments with cells from desensitized animals or from animals which failed to exhibit positive skin test reactions after being submitted to the usual sensitizing procedure. Peritoneal exudate cells from these animals were only insignificantly less inhibited by the specific hapten (potassium dichromate) than cells from hypersensitive guinea pigs (Figure 17). This means that lymphocytes from desensitized animals are still capable of producing detectable amounts of MIF.

It seems that for the inhibition of migration of macrophages, less effector cell activity is required than for eliciting an inflammatory skin reaction. This would mean that the MMT is more sensitive than skin challenges. However, its practical use for diagnostic purposes still has serious technical difficulties.

C. Possible Mechanisms of Tolerance and Desensitization

Initially, we attempted to explain the results of our desensitization and tolerance experiments on the basis of Burnet's clonal selection theory[263] which was, at that time,

102 Chromium: Metabolism and Toxicity

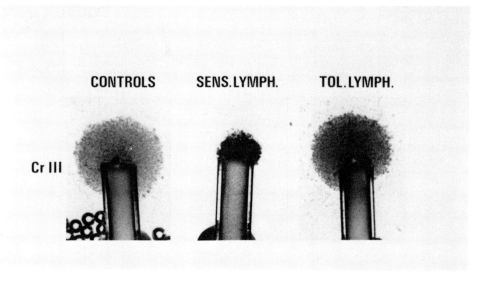

FIGURE 18. Indirect MMT with cells from sensitized and tolerant guinea pigs. A representative experiment.

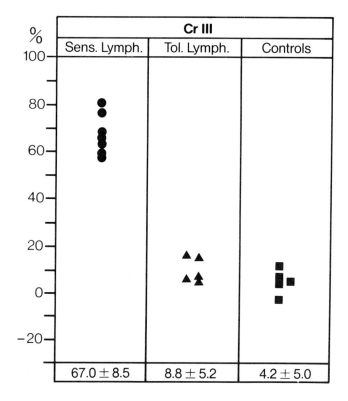

FIGURE 19. Indirect MMT with cells from sensitized and tolerant animals. Results expressed as in Figure 4.

the ruling immunological dogma. According to this theory, the lack of specific immune response in otherwise normally reactive individuals was considered a result of absence of antigen-reactive cells of a given specificity. Therefore, it was assumed that the amounts of intravenously injected haptens such as potassium dichromate or neosalvarsan were capable of destroying or permanently inactivating all immunocompetent cells of a given specificity including their precursors without substantially altering the rest of the immune system. The results of this tolerogenic procedure was a permanent state of unresponsiveness due to elimination of the whole specific alone. Some early experiments of Chase[309] seemed to be in agreement with this hypothesis of tolerance as a result of clonal deletion. At that time, Chase failed to transfer tolerance with lymphoid cells from tolerant donors to normal syngeneic recipients. Moreover, irradiated guinea pigs which were reconstituted with cells from tolerant animals did not exhibit positive skin reactions when submitted to a sensitizing procedure with the specific hapten. On the other hand, tolerance to DNCB-contact sensitivity was terminated by transfer of lymphocytes from DNCB-hypersensitive syngeneic donors.[310] These results could, however, be explained in two ways: either the tolerant individuals do not possess specific antigen-reactive cells, and therefore, lymphocyte suspension from these animals do not reconstitute irradiated recipients, or the tolerant individuals possess suppressor cells capable of suppressing both the induction and manifestation of contact sensitivity.

The latter assumption was supported by the fact that in experiments, transfer of tolerance in both mice and guinea pigs was finally achieved with lymphocyte suspension from tolerant donors[311] or by parabiosis,[143] respectively. Moreover, lymphocytes from normal animals failed to reconstitute tolerant syngeneic recipients. In recent experiments in mice, it has been demonstrated that both mechanisms, namely, clonal deletion and active suppression, could be detected depending on the mode of tolerization.[312] Intravenous injections of the hapten activated suppressor cells, whereas intravenously injected hapten-modified macrophages induced a clonal deletion.

Therefore, both possible mechanisms have to be taken into account when attempts are made to interpret the three immunological phenomena namely hypersensitivity, tolerance and desensitization, and particularly the "double-shot" phenomenon described in the hypersensitivity to neosalvarsan or chromium.

In order to achieve a permanent state of desensitization, two treatments are required: a high i.v. dose of the hapten and a minimal amount of the same hapten applied intradermally or epicutaneously within 24 hr after i.v. injection. The i.v. injection alone produced temporary inhibition. It was then necessary to independently consider the effect of both these treatments in order to find out the mechanism of the "double-shot" phenomenon.

It is known from our studies in contact sensitivity to DNCB in guinea pigs that a hypersensitive animal possesses two types of sensitized lymphocytes: effector cells present mainly in the peripheral circulation and capable of releasing mediators of contact sensitivity skin reactions, and memory cells mainly localized in the lymph nodes and capable of differentiating into new effector cells.[303] It has been reported by Turk and Stone[133] that immunoblasts in the lymph nodes of sensitized guinea pigs are apparently not influenced by an i.v. injection of the specific hapten. An i.v. administration of 500 mg/kg $DNBSO_3$ 4 days after the sensitizing procedure with DNCB (25 $\mu\ell$, 50% solution in acetone) had no effect either on the appearance or on the number of immunoblasts that reached at that time the highest proportion in the regional lymph node.[298] This i.v. injection of the hapten induced a complete though short-lasting unresponsiveness in sensitized animals.

In further experiments it has been demonstrated that lymph node lymphocytes, as opposed to peripheral (peritoneal exudate and blood) lymphocytes, when taken at the time of complete unresponsiveness, are still capable of transferring contact sensitivity to normal syngeneic recipients and to react to a specific in vitro stimulation by an enhanced DNA-synthesis.[227,313]

From all experiments performed in the DNCB-system, it may be concluded that the i.v. application of the hapten impairs the function of peripheral effector lymphocytes without affecting the memory cells in the lymphoid organs. However, it should be mentioned that this impairment is not due to a deletion of the specific clone but rather to a long-lasting, although not permanent, blockade of the specific receptors. A treatment of peripheral effector lymphocytes with trypsin restored their activity, as was demonstrated by a successful transfer of DNCB-contact sensitivity with peritoneal exudate lymphocytes from desensitized guinea pigs treated in vitro with trypsin.[283]

One question remained unsolved: whether the effect of intravenously injected heavy metal salts such as potassium dichromate is only a temporary inactivation of effector cells and not their complete destruction, as postulated by the clonal deletion theory. Since desensitization induced by the i.v. injection alone is, however, only a transient one; the memory cells could at any rate not have been affected. Therefore, in order to explain the permanent desensitization of chromium-sensitive animals, in contrast to the transient one of DNCB-contact-sensitive guinea pigs, the effect and peculiarities of the epidermal challenge with potassium dichromate have to be considered.

An epicutaneous (and also an intradermal) application of the hapten always leads to a stimulation of the antigen-sensitive cells and to their proliferation in the draining lymph node. This occurs independently of whether the hapten was applied for sensitization or for challenge purposes. The proliferating cells are transformed into effector lymphocytes which pass out into the circulation. In normal animals this results in development of sensitivity and in already sensitized guinea pigs, in a booster of the existing hypersensitivity. The consequences of the differentiation and release into the circulation of newly generated effector cells in hypersensitive animals injected intravenously with the hapten shortly before its epidermal application are dependent on the time and on the amount or more precisely, on the half-life and blood level of the hapten used. It is assumed that in already sensitized animals, all effector cells are inactivated by the desensitizing application of the hapten (''first-shot''). However, new effector cells are continuously formed from their unimpaired precursors and restore the contact sensitivity after a certain time. This reconstitution is apparently dependent on the elimination rate of the intravenously injected hapten. As long as a sufficient amount is present in the circulation, the newly formed effector cells are continuously inactivated, and only when the amount falls below the critical level, can the hypersensitivity state become progressively restored. Failure of intravenously injected $DNBSO_3$, beryllium lactate or chromium chloride, and of smaller amounts of potassium dichromate and neosalvarsan to induce a permanent desensitization can be explained by a relatively fast elimination of the intravenously injected hapten from the circulation.

The mechanism of the permanent desensitization occurring in chromium contact-sensitive guinea pigs after the so-called ''double-shot'' procedure could be explained as follows. The i.v. injection of the hapten inactivates the effector lymphocytes, leaving the memory cells unimpaired as was the case in DNCB-system. However, because of its toxicity and capability of penetrating into the cells, potassium dichromate may not only block the receptors of lymphocytes but rather destroy the effector cells or at least cause a permanent blockade. At the same time, the hapten may stimulate the remaining unimpaired memory cells which are then transformed into effector cells and released (or attracted) into the circulation. This release is substantially enhanced and accelerated by the epidermal application of the hapten (''calling effect'' according to the unpublished suggestion of Lefkovits.[304a] As long as sufficient amounts of hapten are present in the circulation, these newly formed effector cells are continuously and permanently inactivated. If the intravenously injected hapten persists long enough in the circulation, the whole clone of specific lymphocytes may be exhausted and eliminated.

It is clear that the "second-shot" (challenge) must follow relatively soon after the "first-shot" (i.v. injection) since all potential effector cells must be formed and released into the circulation before the level of the hapten falls below the critical values. If the "second-shot" is applied too late, the formation of effector cells is extended over a longer period of time and their numbers finally exceed the progressively diminishing amount of circulating hapten and restore the hypersensitivity.

Thus, it seems that the duration of the desensitization is dependent on the relationship between the number of sensitized effector cells and the number of molecules of the hapten present in the circulation. A relatively small amount of the hapten will inactivate or, in the case of chromium hypersensitivity, destroy only the circulating effector cells and therefore cause only temporary desensitization. A larger amount that persists longer in the circulation will continue to impair the function of the newly formed effector cells mobilized into the circulation by the "second shot" and which may lead to a permanent desensitization. This occurs with haptens with a relatively long half-life and with a probably more intensive destructive effect on specific effector lymphocytes. The excision of the hapten application site diminished or abolished the effect of the "second shot". The consequence was that the newly formed effector cells in the lymph nodes are mobilized into the circulation over a longer period of time and therefore cannot anymore be destroyed by the circulating hapten, whose amount at that time falls below the critical level. It is understandable that the intravenously injected hapten and, to a certain extent, even its epicutaneous application also activates suppressor cells. These cells seem to have (in the guinea-pig model) a very moderate effect on already formed effector cells, thus being an additional factor of lower importance in the induction of desensitization. Their main mode of action resides in preventing the formation of memory cells from the virginal antigen-reactive specific T-lymphocytes. The activation of suppressor cells may, however, be the reason why the original level of contact sensitivity is never completely restored after an i.v. injection of the hapten without its epicutaneous application. However, suppressor cells may prevent the formation of new sensitized cells, i.e., memory and effector lymphocytes from their nonsensitized precursors in permanently desensitized animals.

From all that we know, tolerance by pretreatment, as well as tolerance during the primary response, is based on a different mechanism than desensitization. The hapten present in the circulation would, of course, inactivate effector cells, but these cells start to appear in the circulation on day 6 after the sensitizing procedure, i.e., at the time when the level of the hapten falls below an efficient concentration. Therefore, a permanent specific unresponsiveness (tolerance) must be based on a different, more permanent, mechanism, namely, on activation of suppressor cells before effector cells are formed. Suppressor cells activated preferentially by the intravenously injected hapten (tolerance by pretreatment) prevent the development of immunoblasts from their precursors in nonsensitized guinea pigs, as is shown by the lack of proliferating cells in the draining lymph nodes 4 days after the sensitizing procedure in animals rendered specifically tolerant. However, suppressor cells have little or no effect on already formed effector cells.

If the hapten is injected at the time of sensitization (tolerance during the primary response), the suppressor cells activated under these conditions seem, however, to affect both the afferent and the efferent branch of the immune response.

It is postulated that in tolerance during the primary response two mechanisms are at work. Effector cells formed from immunoblasts in the draining lymph node are permanently inactivated by the hapten present in the circulation. Since fewer effector cells are present in these guinea pigs than in already sensitized animals, the amount of the hapten, however low, may still have a partial effect. However, the main tolerogenic effect is performed by suppressor cells which both affect the remaining effector cells and prevent nonsensitized precursor cells to transform into new memory and effector cells. By the combination of both these mechanisms, a complete and permanent unresponsiveness is achieved.

Another, even if a less probable, mechanism has to be considered as an explanation of the ''double-shot'' phenomenon. The epicutaneous challenge (''second shot'') forms in the case of potassium dichromate and other heavy metals, a permanent depot of hapten which inactivates all newly formed effector cells. This phenomenon was described by Felton[314] and Howard[315] as ''masking of tolerance''. This mechanism would explain why an early excision of the hapten application site prevents the establishing of a permanent desensitization. However, the failure of an excision performed later than after 24 hr to prevent the induction of a permanent desensitization and the failure of a depot of hapten induced later than 48 hr after the i.v. injection to contribute to the permanent desensitizing effect, apparently refute this possibility.

In conclusion it is suggested, in analogy to other systems, that tolerance and desensitization in the chromium system are based on different mechanisms. Tolerance is due to activation of suppressor cells which prevent formation of sensitized memory and effector cells. Desensitization is achieved by inactivation of effector cells due to blockade of their receptors. Permanent desensitization requires, besides this blockade, an additional destruction or permanent suppression of effector cells continuously formed from specific memory cells up to the complete elimination or exhaustion. Suppressor cells also activated by the desensitizing procedure may prevent further development of effector cells, but only to a lesser degree and not sufficiently enough to induce a permanent desensitization. However, they may prevent a complete restoration of contact sensitivity after a temporary desensitization.

The permanent suppression of the function of memory cells and the resulting prevention of formation of new effector cells is the crucial point for achieving a permanent desensitization. Immunosuppressive agents may help in this elimination of specific memory cells by affecting preferentially proliferating lymphocytes stimulated by the i.v. injection of the hapten. However, the use of a single agent had no or only a marginal effect on the duration of desensitization. The partial suppression induced by a low dose of intravenously injected hapten was, however, converted into a complete but transient one by the use of a single immunosuppressive drug. It seems therefore that the application of these drugs, acting at different sites of the immune response individually, does not lead to a permanent inhibition of the function of all sensitized cells. For this purpose, a combined application was required.

According to Denman et al.[316] and Mitchison,[317] ALS acts preferentially on the recirculating lymphocytes having only little effect on lymphocytes located in the lymph nodes. The depletion of the paracortical area of the draining lymph node by treatment with ALS, as described by some authors, may therefore be due to preventing T-lymphocytes from entering this area rather than destroying these cells in the lymph node.

Cyclophosphamide (CY) on the other hand, acts on the cells inside the lymph node by inhibiting the proliferation of antigen-stimulated lymphocytes and the formation of immunoblasts (large pyroninophilic cells). A simultaneous administration of both these agents affects the recirculating pool as well as the lymphocytes homing in the lymph nodes. In that way, not only addition, but potentiation of their effects is achieved. If at the same time the susceptibility of sensitized lymphocytes is enhanced by the intravenously injected hapten inducing a proliferative response, the result is a permanent and specific desensitization. A similar potentiating effect of CY and ALS on the desensitizing effect of the i.v. injection of the hapten into DNCB-contact sensitive guinea pigs was also observed.[317a] Beside acting on different sites of the immune response, the potentiating effect of CY on the activity of ALS was visualized in our experiments dealing with skin-graft rejection in guinea pigs.[318]

In these experiments, lymph nodes draining the sites of graft application in guinea pigs, treated with ALS alone or in combination with CY, were investigated by histological and cytological methods. In lymph nodes from animals treated with ALS, the usual depletion of the paracortical area was observed indicating an impairment of the cell-mediated immune response. The areas involved in the humoral response, i.e., the corticomedullary junction, medullary cords, and germinal centers were intact and exhibited rather a proliferative response as a reaction to a heterologous protein (rabbit serum).

In lymph nodes from animals treated with the combination of ALS and CY, an additional depletion of the B-cell area could be observed, thus indicating that the production of antibodies against foreign proteins (rabbit serum) may also be suppressed in the recipient guinea pigs injected with rabbit ALS. Examination of the serum of these guinea pigs showed that there was a substantial reduction indeed in the level of antirabbit serum antibodies in guinea pigs treated with the combination of ALS and CY as compared to the levels detected in animals treated with ALS alone. From these results it may be concluded that CY, by inhibiting the production of antibodies against the heterologous ALS, reduced the elimination rate of this antiserum, thus prolonging its effect on sensitized lymphocytes. This action of CY may well explain, besides its direct effect on sensitized lymphocytes in the lymph node, the prolongation of the desensitization induced by the i.v. injection of the hapten.

The effect of prednisolone could be deduced from its activity on guinea-pig lymphocytes and its effect on ALS. It has been demonstrated that guinea-pig lymphocytes are, in contrast to mouse or rat lymphocytes, largely resistant to the immunosuppressive effect of corticoids.[301] This would explain the failure of prednisolone to enhance the desensitizing effect of the combination of CY and ALS with the intravenously injected hapten. However, the explanation of the paradoxical, i.e., desensitization-decreasing effect, has to be looked for in the influence of corticoids on the activity of ALS. Cortisone administered prior to or simultaneously with ALS inhibits the immunosuppressive effect of this agent on the skin-graft rejection in mice. It is assumed that the mode of action of ALS lies in a mitogenic stimulation (sterile activation) of lymphocytes and that cortisone blocks this mitogenic effect.[302] On the other hand, it has been suggested that ALS inhibits the immune response not by its mitogenic activity, but by a direct cytotoxic effect on circulating lymphocytes.[316,317] The inhibitory effect of prednisolone on ALS could accordingly be explained by a partial block of this cytotoxicity. However, it has also been shown that hydrocortisone may sometimes potentiate the effect of ALS[262] and it is difficult to explain this disagreement of the results.

It may be concluded that the enhancing effect of the combination of immunosuppressive drugs on the desensitizing effect of intravenously injected hapten is rather a complex one, involving the simultaneous action of these drugs on different sites of the immune response (Figure 20) as well as their mutual potentiating interactions.

VIII. SECONDARY PHENOMENA OCCURRING DURING CONTACT SENSITIVITY

It has frequently been observed that sites of previous eczematous lesions, which are clinically completely healed and do not show any traces of inflammation, behave under

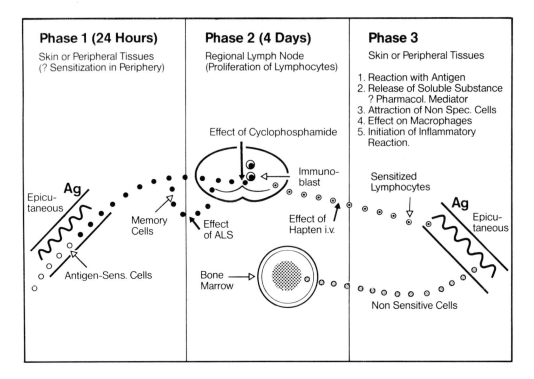

FIGURE 20. Schema of the mechanism of cell-mediated immune reactions and sites of action of different immunosuppressive drugs.

certain conditions differently from the rest of the skin. There exist several reports describing inflammatory reactions of resolved eczematous lesions without a direct contact with the hapten when other parts of the skin were exposed to such a contact or when the hapten was applied orally or intravenously. These inflammatory reactions, which appear spontaneously in already resolved sites of contact sensitivity skin reactions after a remote application of the antigen or the hapten, are known in the clinic under the term "flare-up" reactions.[319-322] A similar phenomenon based probably on the same mechanism is the so-called re-test reaction. It has been shown[323-326] that a completely resolved site of a previous contact sensitivity reaction does react to a new application or exposure to the hapten in an accelerated way and with a higher intensity than the rest of the previously not involved skin.

These observations were also confirmed in experimental animals. Walthard[327] succeeded in eliciting a flare-up reaction of skin sites of already resolved contact sensitivity reactions to nickel sulfate in nickel-hypersensitive guinea pigs by an epicutaneous application of this hapten to the other not previously treated flank of the animals.

Flare-up reactions of previously positive but resolved reaction sites to an intradermal injection of neosalvarsan in guinea pigs with delayed hypersensitivity to this compound were described by Frei,[328] Sulzberger,[265] and Kaplun and Moreinis.[329] These reactions occurred as the result of an i.v. injection of neosalvarsan for desensitizing purposes.

Analogous flare-up reactions were observed by Frey et al.[240] upon an i.v. injection of dinitrobenzenesulfonic acid in about 50% of the guinea pigs highly hypersensitive to dinitrochlorobenzene. An i.v. injection of this hapten (DNBSO$_3$) was also able to elicit an unspecific inflammatory (flare-up) reaction of previously positive skin test sites to another hapten, neosalvarsan, in guinea pigs hypersensitive to the latter arsenic com-

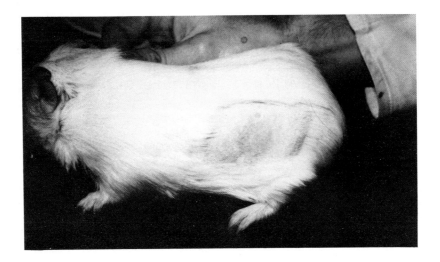

FIGURE 21. Flare-up reaction of a potassium dichromate hypersensitive guinea pig 24 hr after an i.v. injection of 20 mg/kg potassium dichromate.

pound.[330] This unspecific flare-up reaction is based on a different mechanism and may be considered as a local Shwartzman phenomenon.

Another type of flare-up reaction different from those occurring on previously positive resolved lesions are the so-called spontaneous flare-up reactions. These inflammatory reactions are the manifestation of the first appearance of specific effector cells in the circulation and reflect the encounter of these cells with hapten (antigen) remaining in the skin after the sensitizing procedure.[141,331] It is suggested that the hapten is present bound to Langerhans' cells.

A. Secondary Flare-Up Reaction

In the present study, we concentrated our efforts to elucidate the mechanism of secondary flare-up reactions occurring at previous contact sensitivity reaction sites in guinea pigs sensitized to chromium salts. These reactions manifest a striking similarity with the clinically important "fixed drug eruptions". The regularity with which flare-up reactions could be induced by i.v. injections of potassium dichromate into hypersensitive guinea pigs in the course of our desensitization studies enable us to approach this problem experimentally.

Our experiments were performed in albino guinea pigs of the Hartley strain. Sensitization with potassium dichromate and beryllium fluoride and skin challenges with these metal salts were performed as described previously. Only guinea pigs exhibiting a strong contact sensitivity to these metals were used. Flare-up reactions were elicited by i.v. injections of different doses of potassium dichromate or beryllium lactate at different time intervals after the last epicutaneous challenge. In some experiments, chromium hypersensitive guinea pigs were injected intravenously with the trivalent chromium sulfate instead of the hexavalent potassium dichromate.

Flare-up reactions were first seen in some animals as early as 2 hr after i.v. injection of potassium dichromate as a faint pink blush at the site of previous contact reactions (Figure 21). In 4 hr, all animals showed a reaction and in some this was already fully developed. In 6 hr, the flare-up reaction was completely developed in all animals in the highest intensity.

The reaction persisted for 24 hr, though weaker reactions could be seen to be on the

Table 17
PROPORTION OF SENSITIZED ANIMALS SHOWING FLARE-UP OF CONTACT SKIN TEST SITE INDUCED 2 WEEKS PREVIOUSLY AS A FUNCTION OF INTENSITY OF CONTACT REACTION AND THE DOSE OF $K_2Cr_2O_7$ INJECTED INTRAVENOUSLY

Dose of $K_2Cr_2O_7$ (mg/kg)	Intensity of contact reaction 2 weeks previously	Proportion of animals showing flare-up	Proportion of animals with generalized rash
20	0	2/8	0/8
20	+	9/10	0/10
20	++	20/20	6/20
10	++	7/7	0/7
2	++	7/7	0/7
0.1	++	5/7	0/7
0.01	++	0/7	0/7

Table 18
INTENSITY OF FLARE-UP REACTION AS A FUNCTION OF TIME BETWEEN LAST EPICUTANEOUS CONTACT AND INTRAVENOUS INJECTION

Time after last skin test	Intensity of flare-up		
	++	+	0
14 days	100.0%	0	0
1 month	28.5%	57.0%	14.5%
2 months	28.5%	43.0%	28.5%
3 months	14.5%	14.5%	71.0%

wane. Traces of a reaction were still present 48 hr after the i.v. injection but were being resorbed with slight scaling of the epidermis, which lasted up to 1 week after the injection.

In the first set of experiments, the dependence of the flare-up reactions on the dose of the intravenously injected hapten and on the time of this injection has been investigated. The results of these experiments are listed in Tables 17 and 18.

Flare-up reactions were elicited in highly chromium contact-sensitive guinea pigs by an i.v. injection of different doses of the hapten (potassium dichromate) 2 weeks after the last epicutaneous challenge with this chromium compound. At that time, the eczematous skin reaction was completely resolved. It could be demonstrated that the frequency of the flare-up reactions was dependent on the dose of the potassium dichromate injected and on the degree of contact sensitivity of the guinea pig as expressed by the intensity of the last skin reaction. A dose of 20 mg/kg potassium dichromate, which was sufficient to induce a complete and permanent desensitization, elicited a flare-up reaction in all guinea pigs exhibiting strong or weak positive reactions to the epicutaneous challenge. Moreover, from eight animals submitted to the sensitizing procedure with potassium dichromate which failed to react to an epicutaneous challenge with the hapten (skin test negative), two exhibited a flare-up reaction at the site of the last application of 0.5% solution of potassium dichromate, when injected intravenously with 20 mg/kg of this chromium salt. This could be taken as evidence for the existence of a weak hypersensitivity with a subliminal reaction to the challenging dose of the hapten.

A possible existence of a subliminal chromium allergy in man was reported by Folesky[393] and may explain false negative patch test reactions in apparently chromium hyersensitive patients. Lower doses (10 and 2 mg/kg) of intravenously injected potassium dichromate were also capable of inducing a flare-up reaction in all strongly chromium hypersensitive guinea pigs and even a dose of 0.1 mg/kg of intravenously injected chromium salt still induced a flare-up reaction in five of seven strongly chromium-hypersensitive guinea pigs 2 weeks after the last epicutaneous challenge. A dose ten times lower (0.01 mg/kg) failed to elicit a flare-up reaction in previously positive contact sensitivity reaction sites.

A strong flare-up reaction in all hypersensitive guinea pigs was induced by 20 mg/kg potassium dichromate injected 14 days after the last epicutaneous challenge. The frequency of positive flare-up reactions and their intensity progressively decreased, however, as the time interval between the last epicutaneous challenge and the i.v. injection of the hapten was increased from 14 days to 1, 2, or 3 months.

Whereas flare-up reactions, even if of lower intensity, could still be elicited in about three fourths of the animals 1 and 2 months after the last challenge, only one fourth of them reacted with a flare-up reaction to the i.v. injection 3 months after the epicutaneous challenge. Flare-up reactions induced by i.v. injection of potassium dichromate at an earlier time interval than 14 days after the last challenge did not differ in any respect from reactions induced at a 14-day interval.

It is of great interest that flare-up reactions were regularly produced in potassium dichromate-sensitized guinea pigs by an i.v. injection of chromium sulfate in which chromium was present in its trivalent form as opposed to the hexavalent form used for sensitization. This result could be considered a further evidence for the hypothesis that contact sensitivity to chromium is directed against a common determinant independent of whether the metal salts were applied in their tri- or hexavalent forms. Whether this common determinant consists of tri- or hexavalent forms has still to be elucidated.

Intravenous injections of potassium dichromate did not elicit any flare-up reactions either at previously positive contact sensitivity reaction sites to other haptens such a oxazolone or at delayed hypersensitivity reaction sites to tuberculin. No flare-up reactions occurred at the sites of repeated skin painting with 0.5% potassium dichromate in naive (nonsensitized) guinea pigs. These results confirm the immunological specificity of the flare-up reactions studied in our experiments.

In analogy to the frequency of flare-up reactions in dinitrochlorobenzene sensitivity, flare-up reactions at positive contact sensitivity sites were seen in 13 of 23 animals hypersensitive to beryllium fluoride when injected intravenously with 5 mg/kg beryllium lactate 1 week after a positive skin reaction was produced by an epicutaneous application of 0.02 mℓ 1% beryllium fluoride in Triton® X-100.

In order to more precisely characterize the nature of flare-up reactions in chromium-hypersensitive guinea pigs, the skin test sites before and after elicitation of flare-up reaction were examined histologically.

The picture of a positive contact reaction site 14 days after its elicitation is presented in Figure 22. A marked thickening of the epidermis and also a marked infiltrate of mononuclear cells located immediately under the epidermis could still be observed. This finding correlates with the expressed chronicity of the chromium eczematous lesions. The histological picture is considerably changed by the i.v. injection of potassium dichromate (Figure 23). At the peak of the flare-up reaction, i.e., 6 hr after i.v. injections, a dilatation of the capillary network immediately under the epidermis could be observed. The capillaries are packed with polymorphonuclear leukocytes which are also to be seen migrating through the vessel wall into the surrounding tissue.

This histological picture is that of an Arthus phenomenon rather than a contact sen-

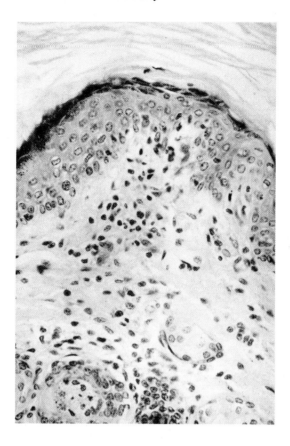

FIGURE 22. Histological picture of guinea pig skin which had been the site of a contact sensitivity reaction 2 weeks previously. Hematoxylin and eosin (140×).

sitivity reaction, because of the polymorphonuclear infiltrate and the early occurrence, and casts doubt on whether the flare-up reactions are in fact manifestations of cell-mediated immunity. However, more experiments should be undertaken to determine the nature of the flare-up reaction and to prove that the mechanism which accounts for the induction of flare-up reaction, when the antigen was administrated intravenously, is different from that which produced the contact sensitivity reaction when the antigen was applied epicutaneously.

The permanent desensitization to chromium might be helpful in this respect. As already mentioned, guinea pigs highly hypersensitive to potassium dichromate became completely and permanently unresponsive to contact sensitivity when injected intravenously with a higher dose of this hapten followed by its epicutaneous application within 24 hr ("double-shot"). Production of a flare-up reaction in such a desensitized animal would be a strong support for the concept that this reaction is induced by a mechanism other than that involved in contact sensitivity.

Guinea pigs exhibiting a high degree of contact sensitivity were injected intravenously with 20 mg/kg of potassium dichromate and challenged epicutaneously with an 0.5% solution of this hapten in 1% Triton® X-100 (double-shot procedure).

As expected, all animals failed to react to this epicutaneous hapten application (second application site). At the same time, the site of the last positive reaction induced 14 days previously (first application site) flared up. A second i.v. injection of 20 mg/kg potassium dichromate 1 week after the first one again elicited a flare-up reaction

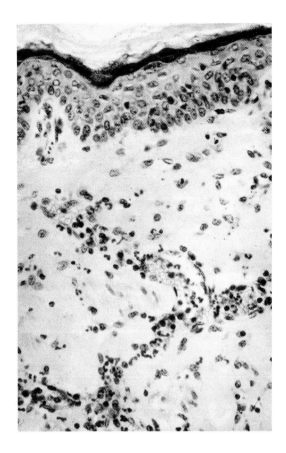

FIGURE 23. Histological picture of a flare-up reaction in a potassium dichromate-hypersensitive guinea pig elicited by an i.v. injection of 20 mg/kg potassium dichromate.

at the first (previously positive) contact sensitivity site. The second hapten application site, where no skin inflammation occurred, failed however, to exhibit a flare-up reaction. An epicutaneous challenge with the same dose of the hapten 24 hr after the second i.v. injection (third site) again failed to produce a positive contact sensitivity reaction. A third i.v. injection of the high dose of potassium dichromate 1 week after the second one again produced a flare-up reaction only at the first hapten application site which had been that of the positive contact sensitivity reaction. The second and the third hapten application sites, which had been those of negative reactions, failed to show any flare-up reactions.

It is evident that the chromium-hypersensitive guinea pigs became completely unresponsive to this metal after the i.v. injection of potassium dichromate, with respect to epicutaneous, i.e., cell-mediated reaction-inducing challenges. The negative results of these challenges were further confirmed by the failure of this hapten application site to exhibit a flare-up reaction upon an i.v. injection of the specific hapten. From this point of view, the second and third hapten application sites resembled the hapten application sites in naive (nonsensitized) animals.

However, the first previously positive hapten application site behaved differently. At that site, a flare-up reaction occurred upon every i.v. injection of the hapten. This could be explained in two ways: either the effector T-cells of contact sensitivity present in the persisting infiltrate in the epidermis are resistant to the desensitizing procedure, or the desensitization concerns only cell-mediated immune reactions leaving the antibody-mediated responses unimpaired. Whereas the first explanation does not seem to be very probable in view of the fact that all specific peripheral lymphocytes are inactivated by the double-shot procedure, the second possibility deserves further consideration. According to this hypothesis, the flare-up reaction develops as a result of the action of antichromium humoral antibodies. Since there was no suppression of the flare-up reaction at positive contact sensitivity sites induced before the desensitizing procedure, it is assumed that the desensitization by the double-shot procedure induces a so-called split tolerance or immune deviation (6), i.e., an immunological tolerance confined to contact or delayed hypersensitivity, while leaving the activity of the humoral antibody-mediated system unimpaired.

In order to differentiate between the immediate- and delayed-type nature of the flare-up reaction in chromium hypersensitivity, the blocking effect of the heterologous antipolymorphonuclear leukocyte serum and anti-lymph node permeability factor (LNPF) serum was investigated. It is known that the first antiserum blocks preferentially the immune reactions of the immediate-type (Arthus reactions) and the latter the delayed-type hypersensitivity reactions. Furthermore, histological examinations were also used for distinguishing between these two types of immune reactions.

The antiserum against guinea-pig polymorphonuclear leukocytes was prepared in rabbits as described by Humphrey.[332] Guinea-pig granulocytes were obtained from exudates induced in the peritoneal cavity by injecting 50 and 20 mℓ of sterile 3% bacteriological peptone with a 15 hr interval. The peritoneal exudate was harvested 3 hr after the last intraperitoneal injections, the cells washed 3 times, and 10^8 granulocytes injected intravenously into rabbits on three occasions with a 7-day interval. The rabbits were bled out 1 month after the last injection. The heat-inactivated (56°C, 30 min) serum was first absorbed with an equal volume of guinea-pig erythrocytes followed by a second absorption with one quarter volume of mixed spleen and mesenteric lymph node cells.

The antiserum against a membrane-free extract of guinea-pig lymph nodes, the so-called anti-LNPF serum, was prepared in rabbits from pooled guinea-pig lymph nodes as described by Willoughby et al.[333] The activity of the serum was assessed by its ability to completely block contact sensitivity to dinitrochlorobenzene. Guinea pigs were sensitized by an epicutaneous application of 20 $\mu\ell$ of 50% dinitrochlorobenzene solution in acetone and challenged 10 days later with 20 $\mu\ell$ of a 0.5% solution of dinitrochlorobenzene in 2-ethoxyethanol. Animals exhibiting a strong confluent red skin reaction with its maximum at 24 hr were injected intravenously with 1 mℓ anti-LNPF serum 24 hr later and challenged epicutaneously with 20 $\mu\ell$ of a 0.5% solution of dinitrochlorobenzene in 2-ethoxyethanol immediately afterwards. Only sera completely inhibiting the skin inflammatory reaction were used in further experiments. All antisera were prepared in New Zealand white rabbits weighing 2 to 3 kg and were sterilized by millipore filtration before use.

For histological examination, skin of flare-up reaction sites was taken from the anesthetized animals and fixed in corrosive acetic fixative (95 parts of a saturated solution of $HgCl_2$ and 5 parts of glacial acetic acid) for 4 hr, then transferred into 70% ethanol. Following this, they were processed in the normal manner, embedded in paraffin and sectioned at 5 μm. The sections were stained with hematoxylin and eosin.

Anti-LNPF serum in a dose of 2 mℓ/kg given intravenously immediately before the i.v. injection of 20 mg/kg potassium dichromate, failed to affect the flare-up of contact sensitivity reaction site induced 14 days previously. This serum was shown to be extremely potent in suppressing delayed hypersensitivity in that it completely abolished contact reactions to dinitrochlorobenzene when injected intravenously immediately before skin painting. This might to some extent exclude the possibility that the flare-up reaction is mediated by a cell-mediated mechanism.

Table 19
EFFECT ON FLARE-UP REACTION OF ANTIPOLYMORPHONUCLEAR SERUM (1 or 2 mℓ) INJECTED INTRAPERITONEALLY 2 DAYS BEFORE 20 mg/kg $K_2Cr_2O_7$ INTRAVENOUSLY

Animal number	Polymorphonuclear leukocytes (per mm^3)	Lymphocytes (per mm^3)	Intensity of flare-up reaction
1	0	3,048	0
2	0	1,023	0
3	0	6,162	0
4	0	2,861	0
5	0	2,150	0
6	0	7,100	0
7	0	1,225	+
8	0	5,148	+
9	445	3,560	0
10	5,445	6,550	++
11	1,440	6,930	+
12	230	1,023	++
Controls			
1	3,749	4,319	++
2	3,190	10,150	++
3	7,128	5,940	++
4	2,205	7,670	++

Table 20
COMPARISON OF DIFFERENT TYPES OF SKIN REACTIONS

	Flare-up reaction	Arthus phenomenon	Contact hypersensitivity reaction
Peak of reaction	6–8 hr	2–4 hr	24–48 hr
Type of cell infiltration	Polynuclear	Polynuclear	Mononuclear
Elicitation in "contact-desensitized" animals	Yes	Not done	No
Effect of ALNPFS	No block	40% decrease	Block
Effect of APNS	Block	Block	No block

In further experiments, chromium hypersensitive guinea pigs received intraperitoneally 2 to 4 mℓ/kg antipolymorphonuclear serum 2 days prior to the i.v. injection of potassium dichromate, which otherwise would induce a flare-up reaction. This procedure caused an agranulocytosis in 8 of 12 animals and in 2 other animals a strong leukocytopenia; the number of a circulating lymphocyte was, however, not essentially changed (Table 19). The flare-up reaction in 7 of 12 guinea pigs thus treated was completely blocked and in other guinea pigs clearly diminished. All the described results strongly indicate that flare-up reactions in chromium hypersensitive guinea pigs are mediated by antibodies rather than by a delayed hypersensitivity mechanism (Table 20). When analyzing the development of flare-up reactions, two possible mechanisms have to be considered:

1. The intravenously injected hapten might induce (in already sensitized animals) a secondary B-cell response with the high level of circulating specific antibodies. This

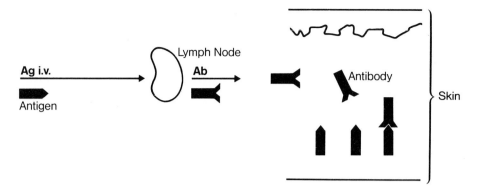

FIGURE 24. Possible mechanism of a flare-up reaction of potassium dichromate hypersensitive guinea pigs induced by an i.v. injection of the hapten. Circulating antibodies induced by the i.v. injection of the hapten reach the residual hapten at the previous hapten application site.

may be possible, since contact sensitivity is usually accompanied by a production of specific antibodies.[122] These circulating antibodies could then react with the antigen residua still present at the site of previous hapten application. It is known that particularly heavy metals remain for an extremely long time in the tissues.[191-193] In such a way, local antigen-antibody complexes are formed which are capable of inducing an Arthus reaction (Figure 24). Results pointing in this direction were published by Pick and Feldmann.[334] These authors have shown that serum of tuberculin-hypersensitive guinea pigs taken 4 hr after an intracardial injection of old tuberculin is capable of transferring cutaneous sensitivity to normal recipients. The skin reaction developing in recipients after intradermal challenging injection of old tuberculin to these animals was considered of an Arthus- or of cutaneous anaphylaxis-type on the basis of histological examinations. The mediator of this skin reaction, which looked fairly similar to the flare-up reaction in our experiments, was γ-immunoglobulin.

2. The second possible mechanism is based on the finding that the cellular infiltrate in chromium contact sensitivity reactions persists in the skin for a relatively long period of time. It is known that this infiltrate contains, besides a small number of specific effector T-cells, different types of mononuclears probably including also B-lymphocytes. Moreover, basophils, known to be sources of several vasoactive mediators, have also been detected in contact sensitivity lesions in a considerable amount (up to 20%).[335] Therefore, it may be assumed that the intravenously injected hapten reaches the skin via blood vessels and capillaries there inducing a local production of specific antibodies (Figure 25). These antibodies may either be bound to the rest of hapten present at the skin reaction site, thus forming antigen-antibody complexes or to basophils. Antibody-coated basophils are capable, upon antigenic stimulation of releasing vasoactive mediators causing the flare-up reaction.

The flaring-up of earlier positive contact-sensitivity skin lesions, but not of negative hapten application sites, in desensitized animals could be explained by assuming that the desensitizing procedure inhibited the activity of only those cells involved in cell-mediated reactions, but not of cells involved in the immediate type. This phenomenon is called split tolerance.

In order to elucidate by which of these two mechanisms the flare-up reaction was mediated, we started to look for circulating antibodies which might be induced by the intravenously injected potassium dichromate and which might react with the residual

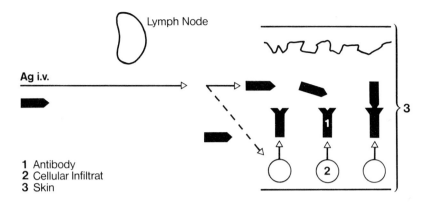

FIGURE 25. Possible mechanism of flare-up reaction of potassium dichromate hypersensitive guinea pigs induced by an i.v. injection of the hapten. The intravenously injected hapten stimulates the lymphocytes present in the residual cellulas infiltrate in the skin. These cells produced local antibodies capable of reacting with the intravenously injected hapten.

hapten still present in the resolved lesions probably in a conjugated form and/or bound to Langerhans' cells. For this purpose, sera from chromium sensitive guinea pigs taken before and after the flare-up-inducing injection of 20 mg/kg potassium dichromate were investigated for the presence of specific antibody. The following methods of antibody detection were used:

1. Double diffusion precipitation in gel (Ouchterlony).
2. Radioimmunoelectrophoresis (212) with ^{51}Cr-labeled chromium conjugated proteins.
3. Passive cutaneous anaphylaxis using potassium dichromate, chromium-conjugated guinea pig serum, or chromium-conjugated skin extracts as eliciting antigens.
4. Attempts were also made to detect antibodies directed against trivalent chromium bound to sheep red blood cells. The method used was direct hemolysis or release of ^{51}Cr from red blood cells coated with radioactive chromium chloride.

However, all our attempts to demonstrate circulating antibodies against chromium conjugated with serum proteins, skin extracts, or erythrocytes failed independently of whether the serum was obtained before or after the i.v. injection of potassium dichromate. These findings excluded, to some extent, the possibility that the flare-up reaction is caused by interaction of circulating antibodies with the residual hapten in the skin. A further argument against this mechanism of flare-up was the observation that a flare-up reaction could only be produced at hapten application sites which have reacted to the epicutaneous challenge and where a cellular infiltrate was still present. Haptenic or antigenic depots at skin areas which had not reacted to the skin challenge were not sufficient of themselves to induce a flare-up reaction following the i.v. injection of potassium dichromate. This applies to the majority of negative skin test sites in insufficiently hypersensitive animals as well as in desensitized animals.

Having failed to prove the first mechanism, we turned our attention to the second one, namely, to the possibility that the intravenously injected hapten, converted into an immunologically active form (antigen) by conjugation with serum or other proteins and/or by processing and presentation by macrophages, may stimulate specific antibody production in lymphocytes present in the remaining cellular infiltrate at the previous contact sensitivity reaction site. The locally produced antibodies could form antigen-antibody complexes with the antigen present on stimulator cells (Langerhans' cells) and/or bound to skin or serum proteins. In this way, the development of an Arthus reaction could be achieved. Another possibility is that the locally produced antibodies are bound to basophils by their Fc-segments and the antigen coming from the circulation may trigger these cells to synthetize and release vasoactive mediators by bridging the antigen-binding sites of the antibodies on the surface of the basophils.

This would induce a cutaneous anaphylactic reaction. However, it is not known whether basophils, certainly present in the acute contact sensitivity lesion, are also capable of persisting in the residual infiltrate. Moreover, the histology of the flare-up reactions resembles more that of the Arthus phenomenon than cutaneous anaphylaxis.

A local antibody production in vitro is a well-established fact in the Jerne plaque technique. However, several authors have discussed the possibility of in vivo local antibody production.

Gell and Hinde[336] and Arnason and Waksman[323] demonstrated plasma cells present several days at skin sites of earlier positive tuberculin reactions. A vigorous local plasmatic response superimposed upon an initial delayed hypersensitivity reaction in the cornea was described by Flax et al.[337] Furthermore, it has been shown that lymphocytes from tuberculin hypersensitive guinea pigs are able to synthetize specific antibodies when injected intradermally to normal recipients to which the antigen (tuberculin) is injected intravenously afterward.[338-340]

In our experiments, a similar protocol was used in order to investigate whether lymphoid cells localized in the skin are capable of producing the type of antibodies required for the induction of the flare-up reaction. Peritoneal exudate cells from either hypersensitive or desensitized donors were injected intradermally into the skin of normal recipients in order to produce a mononuclear infiltrate. An attempt was made to elicit a flare-up reaction at the site of this artificially induced infiltrate by injecting the antigen intravenously. This experiment was performed using six highly potassium dichromate hypersensitive, four hypersensitive but specifically desensitized, and six normal guinea pigs. The permanent desensitization to chromium contact sensitivity was achieved by i.v. injection of a high dose of potassium dichromate followed by an epicutaneous application of this hapten within 24 hr.

The peritoneal exudate was induced by an intraperitoneal injection of 30 mℓ sterile light paraffin oil and cells harvested 4 days later by washing out the peritoneal cavity of sacrificed animals with Hanks' BSS. The exudate cells consisting of about 69% macrophages, 14% lymphocytes, and 17% granulocytes were washed, resuspended in Hanks' BSS, and 0.3 mℓ of the cell suspension containing about 3×10^7 cells (4×10^6 lymphocytes) were injected intradermally into six normal guinea pigs. Since the recipients were not syngeneic, a transient normal lymphocyte transfer reaction occurred at the intradermal injection site. When the nonspecific skin reaction disappeared, 3 to 4 days later, recipients were injected intravenously with 20 mg/kg potassium dichromate. An inflammatory reaction started to develop 6 hr after this i.v. injection, at the site of the intradermal infiltrate of cells from hypersensitive donors, and the peak intensity was observed at 24 hr (Table 21). Similar reactions were observed at sites of injections of cells from desensitized but not from normal donors.

The possibility that the flare-up reaction was a consequence of the normal lymphocyte transfer reaction rather than of the specific lymphocyte-hapten interaction could be excluded on the basis that intradermally injected peritoneal exudate cells from normal donors, which also induced a transient inflammatory reaction, did not produce a flare-up reaction upon the i.v. application of the hapten. However, sensitized lymphoid cells could be nonspecifically stimulated when transferred into an allogeneic environment.[341] An important result is that peritoneal exudate cells from chromium-hypersensitive

Table 21
TRANSFER OF FLARE-UP REACTION

Origin of transferred cells	Number of animals	Erythema	Diameter (mm)
Normal animals	6	+	5
Cr-sensitized animals	6	+++	14
Desensitized animals	4	+++	13

guinea pigs, made permanently unresponsive by the "double-shot" procedure, were still capable of producing a flare-up reaction when injected intradermally into normal recipients. On the other hand, peritoneal exudate cells from desensitized guinea pigs (at least in the DNCB-system)[283] are unable to transfer contact sensitivity. It has been shown already that permanently desensitized chromium hypersensitive guinea pigs are capable of exhibiting flare-up reactions of resolved positive skin reaction sites induced before desensitization, but not of negative sites where the hapten was applied after the desensitizing procedure. This fact, together with the above reported results, strongly suggests that flare-up reactions induced at the site of intradermal injections of peritoneal exudate cells from chromium hypersensitive donors desensitized by the double-shot procedure is mediated by a mechanism different from that involved in contact sensitivity and particularly in local passive transfer of contact sensitivity.[342,343] This is even more so, since local passive transfer is not, in contrast to tuberculin hypersensitivity, feasible in the contact sensitivity system,[344] unless expanded specific T cell lines are used. It seems that flare-up reactions occurring in contact sensitivity lesions in animals sensitized, challenged, and intravenously injected with other antigens, e.g., DNCB, are based on a different mechanism. Several authors have shown[345-347] that in this system flare-up reactions are accompanied by a mononuclear cell infiltrate as opposed to the polynuclear one observed in chromium hypersensitive flare-up reaction. This finding, together with the failure to prevent the development of the flare-up reaction with various antihistamines, suggests that in the DNCB-system flare-up reactions are induced by cell-mediated mechanisms. In view of the different behavior of organic compounds such as DNCB and metal salts (e.g., chromium or arsenic) in the body, this difference is not surprising. It is known that chromium, arsenic, and other heavy metals remain in the tissues for an extremely long period of time,[191-193] whereas other simple chemicals, such as DNCB, are eliminated in a relatively short time. This may to a certain extent explain the differences in the mechanisms of flare-up in these two hypersensitivity systems.

Whereas some authors suggested that in the DNCB-system in guinea pigs the flare-up reaction results from an interaction of sensitized effector T-cells present in the skin infiltrate with the intravenously injected hapten[348] which reached the epidermis conjugated to serum proteins,[349] Jansen and Bleumink[345,346] were of the opinion that in addition to the interaction of sensitized lymphocytes with antigen, serum proteins play an important role by themselves. These authors postulated that the interaction of those three components leads to a denaturation of the proteins, which in turn induces a release of pharmacodynamic mediators. This hypothesis was supported by the fact that in their experiments 10 mg/kg potassium dichromate was capable of inducing a flare-up reaction of previous contact sensitivity lesions in DNCB-sensitized guinea pigs, i.e., in an unspecific way.

However, in the experiments described in this study, flare-up reaction was induced also with very low doses of potassium dichromate (0.1 mg/kg) which have an only negligible denaturating effect. Moreover, the i.v. injection of potassium dichromate induced a flare-up reaction of skin sites which had been earlier positive to chromium, but not of skin sites which had been positive to oxazolon or tuberculin. These results confirm the specificity of the flare-up reactions in the chromium system observed in our experiments. The previously mentioned results—that flare-up reactions could also be produced at earlier positive skin test sites of desensitized animals and that peritoneal exudate cells from these animals are capable of transferring the capacity for flaring-up to normal recipients—strongly suggest that in the chromium system, in contrast to the DNCB-system, the mechanisms of the flare-up reaction is related to that of the Arthus phenomenon. It is known from the literature that besides the secondary flare-up reactions described by several authors,[170,269,345–347,349,350] which appear at previously positive skin reaction sites after a remote hapten application (epicutaneous or systemic), there exists another type of flare-up: the so-called spontaneous one. This reaction, first described by Sulzberger[265] and later confirmed by many other authors,[331,351,352] appears during the development of contact or delayed hypersensitivity at sites where the hapten was applied to naive individuals for the first time for sensitizing purpose. This flaring-up, which is regarded as the reflection of the first encounter of the generated effector cells with the residual antigen at its application site, was observed in delayed hypersensitivity to neosalvarsan,[265] to oxazolon[352a] or to DNCB.[141,352] Histological studies revealed in guinea pigs and in man a prevalently mononuclear infiltrate confirming the cell-mediated character of this phenomenon. Chromium-hypersensitivity spontaneous flare-up reactions have never been reported, which does not exclude their existence.

When animals that were hypersensitive to neosalvarsan were injected intravenously with this compound immediately or not later than 24 hr after the last intradermal challenge, a different flare-up reaction developed at the hapten application site.[330] This reaction appeared after 2 hr, was hemorrhagic in character, and could be also produced in a nonspecific way, e.g., by an i.v. injection of dinitrobenzenesulfonic acid. The mechanism of this reaction, interpreted as a local Shwartzman phenomenon, may be identical with that of a nonspecific flare-up induced in DNCB-sensitive guinea pigs by a high dose of potassium dichromate.[345] In chromium-hypersensitive guinea pigs this type of flare-up was not observed.

Although we have not been able to give direct evidence for the local antibody production by the cells present in the persisting cellular infiltrate in chromium-hypersensitive individuals, the indirect evidence as reviewed in this study seems to be sufficient for classifying the flare-up reaction in the chromium system as an Arthus-like reaction dependent on specific antibodies.

The use of flare-up reaction for diagnostic purposes was reported by Schleiff.[80] An oral administration of 1 to 10 mg/kg potassium dichromate to chromium hypersensitive patients induced a flare-up of previously negative skin challenge sites and of resolved eczematous foci. By this method a latent state of chromium hypersensitivity could be detected. A similar observation was reported by Kaaber.[81]

B. Generalized Rash

When chromium contact-sensitive guinea pigs were injected intravenously with a high (20 mg/kg) dose of potassium dichromate, about 30% of the animals developed a generalized erythematous eruption. This rash did not occur in animals exhibiting a moderate degree of hypersensitivity and could not be elicited with lower doses of the hapten. Macroscopically, the rash was a flat erythematous eruption, though in a few animals it had a definite papular appearance. The generalized rash started as faint pink spots 6 to 8 hr after the i.v. injection and became a definite red eruption by 24 hr. This reaction persisted unchanged for another 24 hr and resolved with scaling over the next

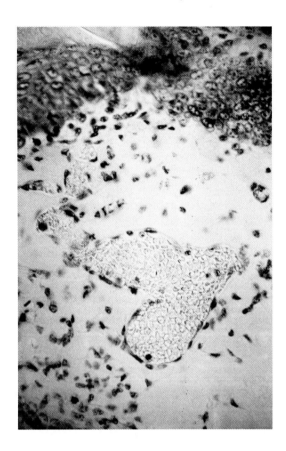

FIGURE 26. Histological picture of a skin rash induced by an i.v. injection of potassium dichromate in animals highly hypersensitive to this hapten.

week. Histologically, the generalized rash at its peak was characterized by a moderate thickening of the epidermis and a marked dilatation of superficial capillaries in the upper dermis. A cellular infiltrate could hardly be observed and there was no perivascular infiltration. In this respect, there was some marked similarity with the skin eruptions observed in guinea pigs with serum sickness (Figure 26).

A generalized rash could not be observed in beryllium fluoride-sensitive guinea pigs injected intravenously with the highest tolerated dose of beryllium lactate. On the other hand, generalized rash was described in guinea pigs sensitized to neosalvarsan,[265,328] picryl chloride,[289] or dinitrochlorobenzene[269,353,354] after an i.v. injection of the respective hapten.

The secondary generalized rashes in human, called "id" reactions and described by Shelmire[355] and Kligman[356] in the course of oral desensitization to poison ivy, exhibit a similar picture and may be caused by similar mechanisms. It has been postulated that the generalized rash can also be produced by absorption of the sensitizer from the skin into the circulation which may explain its clinical occurrence.

As to the pathogenetic mechanism, the interpretation of histology is of great importance. Since the generalized rash is caused by dilatation of superficial capillaries without any perivascular infiltration, it resembles the histological picture of serum sickness. It is believed that this phenomenon is caused by circulating immune complexes which in turn cause local release of pharmacological agents such as vasoactive kinines. Small amounts of bradykinine are known to cause dilatation of capillaries without a perivascular cellular infiltrate.[357]

It may well be that the generalized rashes observed in chromium hypersensitivity are caused by a similar mechanism, namely, by immune complexes formed after the i.v. injection of the hapten. The origin of specific antibodies remains, however, in view of our negative results, unexplained.

The denaturation of proteins by the high dose of hapten, as postulated by Janson and Bleumink,[345] cannot be excluded as the cause of the generalized rash, since these secondary reactions—in contrast to local flare-up reactions—are induced only by very high doses (20 mg/kg) of potassium dichromate. Nevertheless, this interpretation does not seem to be very likely, because the generalized rash is also antigen-specific and does not occur in non- or weakly sensitized guinea pigs.

The massive presence of eosinophils in the generalized rash eruptions in DNCB-sensitized guinea pigs[353,354] leads to the conclusion that generalized rash is, in contrast to contact sensitivity, an expression of type I or III reactions according to the classification of Coombs and Gell.[358] It has been shown that eosinophils are attracted by a chemotactic factor for eosinophils, which is elaborated by an interaction of antibody with antigen.[359] This may occur even in guinea pigs with delayed hypersensitivity.

Another explanation of the generalized rash was stressed by de Weck.[360] He suggested that the generalized rash may be a consequence of a massive synthesis and release of inflammatory lymphokines into the circulation. There are three facts speaking in favor of this interpretation:

1. Guinea pigs with high levels of specific antibodies, but not showing delayed hypersensitivity, do not produce the rash.
2. Hapten-protein conjugates form antigen-antibody complexes but do not elicit the rash.
3. The rash is accompanied by a rise of body temperature as is observed also in delayed hypersensitivity reactions.[360a]

Summarizing the available data about secondary phenomena which accompanied delayed and contact sensitivity reactions, an apparent diversity of mechanisms dependent on the type of hapten or antigen used for sensitization became evident.

IX. CONCLUSIONS

The data reviewed in this study could be summarized in the following points:

1. Chromium eczema is the most frequently occurring occupational dermatosis whose importance is still rising with the increasing use of this metal in the industry and in the manufacturing of household articles.
2. Contact sensitivity to chromium can be best revealed by challenges with hexavalent chromium compounds. Skin challenges with tri- or divalent chromium salts call for higher concentrations of the hapten and/or higher degree of contact sensitivity.
3. In experimental animals (guinea pigs), contact sensitivity to some metals, including chromium, is under genetic control.
4. Both tri- and hexavalent chromium salts are (in guinea pigs) successful sensitizers, but hexavalent compounds are better in vivo elicitors than trivalent ones. The contradiction between the poor protein binding capacity of hexavalent chromium salts and their strong immunogenicity could be explained by the conversion of hexavalent chromium into trivalent by the sulfur-containing amino acids in the skin. Trivalent

chromium binds proteins very efficiently, but cannot penetrate the skin and membranes.
5. In in vitro experiments, it has been demonstrated that hypersensitivity to chromium is directed against one common determinant. It is very likely that in this determinant, chromium is present in its trivalent form.
6. In chromium hypersensitive guinea pigs, a permanent and complete desensitization was achieved by an i.v. injection of high doses of potassium dichromate followed by an epicutaneous challenge with this hapten within 24 hr ("double-shot"). Lower i.v. doses of the hapten or a delay in its epicutaneous application induced an only temporary inhibition.
7. The additional application of a combination of cyclophosphamide (CY) and ALS converted the temporary inhibition achieved by a low i.v. dose of the hapten into a very long-lasting, even though not permanent, desensitization.
8. Intravenous injections of the hapten prior to or at the time of sensitization induced a complete and permanent hapten-specific tolerance. For this purpose, lower doses were required than for a permanent desensitization.
9. Flare-up reactions of resolved positive contact sensitivity reaction sites induced by the intravenously injected hapten are mediated by mechanisms similar to that of Arthus phenomenon, whereas generalized rashes are more close to picture of serum sickness.

In conclusion, it is difficult to predict whether this experimental approach will lead to an improvement of the still unsatisfactory therapy of allergic eczema in general and of chromium eczema in particular. However, to my opinion there is at present no other way than by extending the knowledge about pathogenetical mechanisms involved in this disease and their manipulation, and the only method to achieve this is the experiment.

REFERENCES

1. **Borelli, S. and Düngemann, H.**, Aktuelle Kontaktekzemursachen in der Metallindustrie, *Berufs-Dermatosen*, 12, 1, 1964.
2. **Engelhardt, A. W.**, Das allergische Kontaktekzem in seiner Bedeutung für Praxis und Klinik, *Muench. Med. Wochenschr.*, 102, 1849, 1960.
3. **Feuerman, E. J.**, Housewive's eczema and the role of chromates, *Acta Derm.-Venereol.*, 49, 288, 1969.
4. **Gaul, L. E.**, Metal sensitivity in eczema of the hands, *Ann. Allergy*, 11, 758, 1953.
5. **Gaul, L. E.**, Incidence of sensitivity to chromium, nickel, gold, silver and copper compared to reactions to their aqueous salts including cobalt sulfate, *Ann. Allergy*, 12, 429, 1954.
6. **Lindemayr, W. and Lofferer, O.**, Allergische Berufsdermatosen in Oesterreich, *Wien. Klin. Wochenschr.*, 77, 905, 1965.
7. **Schneider, W. and Wagner, H.**, Kontaktdermatitis, in *Dermatologie und Venerologie*, Vol. 3, Gottron, H. A. and Schönfeld, W., Eds., Georg Thieme, Stuttgart, 1959.
8. **Schulz, K. H.**, Berufsdermatosen, in *Dermatologie und Venerologie*, Vol. 5, Gottron, H. A. and Schönfeld, W., Eds., Georg Thieme, Stuttgart, 1963.
9. **Wagner, G. and Wezel, G.**, Art und Häufigkeit hautschädigender Berufsnoxen in Schleswig-Holstein, *Berufs-Dermatosen*, 14, 1, 1966.
10. **Zelger, J. and Pretner, A.**, Eigene Beobachtungen bei ekzematogenen Chromatexpositionen, *Wien. Klin. Wochenschr.*, 76, 552, 1964.
11. **Fregert, S.**, Occupational dermatitis in a 10-year material, *Contact Dermatitis*, 1, 96, 1975.
12. **Burrows, D.**, Chromium and the skin, *Br. J. Dermatol.*, 99, 587, 1978.

13. **Geiser, J. D.**, Les facteurs de sensibilisation dans l'eczéma au ciment, *Schweiz. Med. Wochenschr.*, 98, 1193, 1968.
14. **Spier, H. W. and Natzel, R.**, Zur Pathogenese des Zementekzems. I. Zementekzem und Chromatallergie, *Arch. Dermatol., Syph.*, 193, 537, 1952.
15. **Cronin, E.**, Contact dermatitis, XV. Chromate dermatitis in men. *Br. J. Dermatol.*, 85, 95, 1971.
16. **Parkhurst, H. J.**, Dermatosis industrialis in a blueprint worker due to chromium compounds, *Arch. Dermotol. Syphilol. (Chicago)*, 12, 253, 1925.
17. **Jadassohn, J.**, Zur Kenntnis der medikamentösen Dermatosen, in *Verhandlungen der Deutschen Dermatologischen Gesellschaft, 5. Kongress, Graz, 1895*, Braunmüller, W., Ed., Vienna, 1896, 103.
18. **Burckhardt, W.**, Experimentelle Sensibilisierung des Meerschweinchens gegen Terpentinöl, *Acta Derm.-Venereol.*, 19, 359, 1938.
19. **Jäger, H. and Pelloni, E.**, Tests épicutanés aux bichromates, positifs dans l'eczéma au ciment, *Dermatologica*, 100, 207, 1950.
20. **Berger, W. and Hansen, K.**, *Allergie*, Georg Thieme, Leipzig, 1940.
21. **Bonnevie, P.**, *Aetiologie und Pathogenese der Ekzemkrankheiten*, A. Busck, Copenhagen and J. A. Barth, Leipzig, 1939.
22. **Schwartz, L., Tulipan, L., and Peck, S.**, *Occupational Diseases of the Skin*, 2nd ed., Lea & Febiger, Philadelphia, 1947.
23. **Urbach, E.**, *Klinik und Therapie der allergischen Krankheiten*, W. Maudrich, Vienna, 1935.
24. **Fregert, S. and Gruvberger, B.**, Chemical properties of cement, *Berufs-Dermatosen*, 20, 238, 1972.
25. **Pirilä, V. and Kajanne, H.**, Sensitization to cobalt and nickel in cement eczema, *Acta Derm.-Venereol.*, 45, 9, 1965.
26. **Rabeau, H. and Ukrainczyk, V.**, Dermites des blanchisseuses. Rôle du chrome et du chlore (en France), *Ann. Dermatol. Syphiligr. (Paris)*, 10, 656, 1939.
27. **Schreus, H. Th. and Bürk, H.**, Ueber die Häufigkeit der Chromsäureallergie in verschiedenen Berufen, *Arch. Gewerbepathol. Gewerbehyg.*, 12, 218, 1944.
28. **Jansen, L. H. and Berrens, L.**, Sensitization and partial desensitization of guinea-pigs to trivalent and hexavalent chromium, *Dermatologica*, 137, 65, 1968.
29. **Macher, E. and Chase, M. W.**, Studies on the sensitization of animals with simple chemical compounds. XI. The fate of labeled picryl chloride and DNCB after sensitizing injections. XII. The influence of excision of allergenic depots on onset of delayed hypersensitivity and tolerance, *J. Exp. Med.*, 129, 81, 1969.
30. **Walsh, E. N.**, Chromate hazards in industry, *JAMA*, 153, 1305, 1953.
31. **Royle, H.**, Toxicity of chromic acid in the chromium plating industry (2), *Environ. Res.*, 10, 141, 1975.
32. **Amrein, H. P. and Miescher, G.**, Zur Aetiologie des Zementekzems, *Z. Unfallmed. Berufskr.*, 48, 140, 1955.
33. **Lejhancová, M. and Wolf, J.**, Die Bedeutung des Kaliumbichromattestes beim Alkaliekzem, *Dermatol. Wochenschr.*, 132, 1273, 1955.
34. **Miescher, G., Amrein, H. P., and Leder, M.**, Chromatüberempfindlichkeit und Zementekzem, *Dermatologica*, 110, 266, 1955.
35. **Skog, E. and Wahlberg, J. E.**, Patch testing with potassium dichromate in different vehicles, *Arch. Dermatol.*, 99, 697, 1979.
36. **Spier, H. W. and Natzel, R.**, Chromatallergie und Zementekzem: Gewerbedermatologischer und analytischer Beitrag, *Hautarzt*, 4, 63, 1953.
37. **Zelger, J.**, Zur Klinik und Pathogenese des Chromatekzems, *Arch. Klin. Exp. Dermatol.*, 218, 499, 1964.
38. **Bohnenstengel, G.**, Kaliumbichromat-Testungen in verschiedenen pH-Bereichen, *Dermatol. Wochenschr.*, 51, 795, 1965.
39. **Cairns, R. J. and Calnan, C. D.**, Green tattoo reactions associated with cement dermatitis, *Br. J. Dermatol.*, 74, 288, 1966.
40. **Calnan, C. D.**, Cement dermatitis, *J. Occup. Med.*, 2, 15, 1960.
41. **Denton, C., Keenan, R. G., and Birmingham, D. J.**, Content of cement and its significance in cement dermatitis, *J. Invest. Dermatol.*, 23, 189, 1954.
42. **Pautritzel, R., Rivasseau, J., and Rivasseau-Coutant, A.**, Sensibilisation aux ions métalliques et maladies professionnelles, *Arch. Mal. Prof. Med. Trav. Secur. Soc.*, 19, 109, 1958.
43. **Pautritzel, R., Rivasseau, J., and Rivasseau-Coutant, A.**, Le traitment des allergies cutanées de contact par epidermodésensibilisation, *Med. Hyg.*, 533, 47, 1962.

44. **Zelger, J. and Wachter, H.**, Ueber die Beziehungen zwischen Chromat- und Dichromatallergie, *Dermatologica*, 132, 45, 1966.
45. **Everall, J. D., Truter, M. R., and Truter, E. V.**, Epidermal sensitivity to chromium, cobalt and nickel, *Acta Derm.-Venereol.*, 34, 447, 1954.
46. **Eberhartinger, C. and Ebner, H.**, Beitrag zur Kenntnis der Chromatkontaktallergie. I. Mitteilung, *Berufs-Dermatosen*, 16, 1, 1968.
47. **Eberhartinger, C. and Ebner, H.**, Beitrag zur Kenntnis der Chromatkontaktallergie. II. Mitteilung, *Berufs-Dermatosen*, 16, 97, 1968.
48. **Eberhartinger, C. and Ebner, H.**, Beitrag zur Kenntnis der Chromatkontakallergie. III. Mitteilung, *Berufs-Dermatosen*, 16, 247, 1968.
49. **Paschoud, J. M.**, Kritische Bemerkungen zur Zusammensetzung der sogenannten Standardreihen für epikutane Testproben, *Dermatologica*, 124, 196, 1962.
50. **Skog, E. and Thyresson, N.**, The occupational significance of some common contact-allergens, *Acta Derm.-Venereol.*, 33, 65, 1953.
51. **Wagner, G.**, Eine Analyse von 500 allergischen Kontaktekzemen, *Berufs-Dermatosen*, 7, 307, 1959.
52. **Brun, R. M.**, Kontaktekezeme durch "unlösliche" Chromatfarben, *Berufs-Dermatosen*, 13, 182, 1966.
53. **Behrbohm, P.**, Allergische Kontaktekzeme durch chromhaltige Rostschutzmittel, *Berufs-Dermatosen*, 5, 271, 1957.
54. **Korting, G. W. and Wiehl, R.**, Ueber Hautschäden durch Holzschutzmittel (Pentachlorphenol), *Z. Haut-Geschlechtskr.*, 14, 280, 1953.
55. **Schwarzkopf, K.**, Ueber Hautschäden durch Holzschutzmittel, *Z. Haut-Geschlechtskr.*, 16, 6, 1954.
56. **Samitz, M. H. and Gross, S.**, Effects of hexavalent and trivalent chromium compounds on the skin, *Arch. Dermatol.*, 84, 404, 1960.
57. **Valér, M. and Somogyi, S.**, Beitrag zur Frage der Chromallergie, *Hautarzt*, 17, 351, 1966.
58. **Kroepfli, H. and Schuppli, R.**, Beiträge zum Waschmittelekzem, *Dermatologica*, 110, 1, 1955.
59. **Krüger, H. and Dorn, H.**, Klinische Beobachtungen zum Waschmittelekzem in Verbindung mit Chrom- und Nickelallergie, *Z. Haut-Geschlechtskr.*, 20, 307, 1956.
60. **Foussereau, J. and Laugier, P.**, Allergic eczemas from metallic foreign bodies, *Trans. St. John's Hosp. Dermatol. Soc.*, 52, 220, 1966.
61. **Fregert, S., Gruvberger, B., and Mitchell, J. C.**, Chromate in postage stamps, *Contact Dermatitis*, 1, 328, 1975.
62. **Stevenson, C. J.**, Fluorescence as a due to contamination in TV workers, *Contact Dermatitis*, 1, 242, 1975.
63. **Krook, G., Fregert, S., and Gruvberger, B.**, Chromate and cobalt eczema due to magnetic tapes, *Contact Dermatitis*, 3, 60, 1977.
64. **Björnberg, A.**, Allergic reactions to chrome in green tattoo markings, *Acta Derm.-Venereol.*, 39, 23, 1959.
65. **Björnstad, R.**, Idiosyncratic eczema with eruptions also in old tattooing, *Acta Derm.-Venereol.*, 30, 379, 1950.
66. **Heilesen, B.**, Chromium allergy as the probable cause of eczema in tattooing, *Acta Derm.-Venereol.*, 33, 255, 1953.
67. **Loewenthal, L. J. A.**, Reaction in green tattoos, *Arch. Dermatol.*, 82, 237, 1960.
68. **van Tongeren, J. M. H. and Majoor, C. L. H.**, Demonstration of protein-losing gastroenteropathy. The disappearance rate of 51 Cr from plasma and the binding of 51 Cr to different serum proteins, *Clin. Chim. Acta*, 14, 31, 1966.
69. **Jacobs, F.**, Ueber Epikutan- und Schleimhauttestungen, *Dermatol. Wochenschr.*, 127, 446, 1953.
70. **Tilsley, D. A. and Rotstein, H.**, Sensitivity caused by internal exposure to nickel, chrome and cobalt, *Contact Dermatitis*, 6, 175, 1980.
71. **Polak, L., Turk, J. L., and Frey, J. R.**, Studies on contact hypersensitivity to chromium compounds, in *Progress in Allergy*, Vol. 17, Kallós, P., Waksman, B. H., and de Weck, A., Eds., S. Karger, Basel, 1973, 145.
72. **Merritt, K. and Brown, S. A.**, Hypersensitivity to metallic biomaterials, in *Systemic Aspects of Biocompatibility*, Williams, D. F., Ed., CRC Press, Boca Raton, Fla., 1979, chap. 21.
73. **Schwarz-Speck, M. and Keil, H.**, Experimentelles Chrom-III-Ekzem, *Dermatologica*, 130, 373, 1965.
74. **Geiser, J. D., Jeanneret, J. P., and Delacrétaz, J.**, Eczéma au ciment et sensibilistation au cobalt, *Dermatologica*, 121, 1, 1960.
75. **Storck, H. and Schwarz, M.**, Observations on allergenicity of simple inorganic compounds, *Acta Allergol.*, (Suppl. 7), 232, 1960.

76. **Marson, G.,** Studio della reattività cutanea alle intradermoreazioni con bicromato di potassio in pazienti affetti da eczema da cemento, *Minerva Dermatol.,* 32, 282, 1957.
77. **Wahlberg, J. E.,** Thresholds of sensitivity in metal contact allergy. I. Isolated and simultaneous allergy to chromium, cobalt, mercury and/or nickel, *Berufs-Dermatosen,* 21, 22, 1973.
78. **Wahlberg, J. E.,** Thresholds of sensitivity in metal contact allergy. 2. The value of percutaneous absorption studies for selection of the most suitable vehicle, *Berufs-Dermatosen,* 21, 151, 1973.
79. **Polak, L. and Turk, J. L.,** Studies on the effect of systematic administration of sensitizers in guinea-pigs with contact sensitivity to inorganic compounds. II. The flare-up of previous test sites of contact sensitivity and the development of a generalized rash, *Clin. Exp. Immunol.,* 3, 253, 1968.
80. **Schleiff, P.,** Provokation des Chromatekzems zu Testzwecken durch interne Chromzufuhr, *Hautarzt,* 19, 209, 1968.
81. **Kaaber, K. and Veien, N. K.,** The significance of chromate ingestion in patients allergic to chromate, *Acta Derm.-Venereol.,* 57, 321, 1977.
82. **Bandmann, H. J. and Fuchs, G.,** Ueber die Kobaltkontakallergie, ihre Beziehung zur Bichromat- und Nickelkontaktallergie, sowie ihre gewerbedermatologische Bedeutung, *Hautarzt,* 14, 207, 1963.
83. **Hilt, G.,** Hexavalent chromate dermatitis in the group of exzematous dermatitis caused by sensitization to metals, *Dermatologica,* 109, 143, 1954.
84. **Meneghini, C. L.,** Beitrag zum Studium der Berufsdermatosen durch Zement und Kalk, *G. Ital. Dermatol.,* 93, 303, 1952.
85. **Nater, J. P.,** Cementeczeem, *Ned. Tijdschr. Geneeskd.,* 102, 250, 1959.
86. **Pirilä, V.,** On the role of chrome and other trace elements in cement eczema, *Acta Derm.-Venereol.,* 34, 136, 1954.
87. **Zelger, J.,** Zur Kenntnis der Kobalt-Allergie, *Derm. Wochenschr.,* 146, 425, 1962.
88. **Zina, G.,** Cromoreattività cutanea e dermatosi professionali, *Minerva Dermatol.,* 31, 305, 1956.
89. **Zina, G. and Bonu, G.,** Un tema di reattività cutanea al cromo, al nickel e al cobalto. Reattività crociata e risposte cosiddette tardive tubercoliniche, *Minerva Dermatol.,* 32, 56, 1957.
90. **Fregert, S. and Rorsman, H.,** Allergy to chromium, nickel and cobalt, *Acta Derm.-Venereol.,* 46, 144, 1966.
91. **Hilt, G.,** Cosensibilisation et polysensibilisation dans la dermatite du chrome hexavalent, *Bull. Soc. Fr. Dermatol. Syphiligr.,* 63, 512, 1957.
92. **Mikulecký, Z.,** Příspěvek k otázce sdružené alergie na chrom-nikl-kobalt, *Cesk. Dermatol.,* 38, 284, 1963.
93. **Pirilä, V. and Kajanne, H.,** Sensitization to cobalt and nickel in cement eczema, *Acta Derm.-Venereol.,* 45, 9, 1965.
94. **Mitchell, J. C.,** The angry back syndrome: eczema creates eczema, *Contact Dermatitis,* 1, 193, 1975.
95. **Mitchell, J. C.,** Multiple concomitant positive patch test reactions, *Contact Dermatitis,* 3, 315, 1977.
96. **Landsteiner, K. and Jacobs, E.,** Studies on the sensitization of animals with simple chemical compounds, *J. Exp. Med.,* 61, 643, 1935.
97. **Engelhardt, W. B. and Mayer, R. J.,** Das Chromekzem im graphischen Gewerbe, *Arch. Gewerbepathol. Gewerbehyg.,* 2, 140, 1931.
98. **Epstein, S.,** Contact dermatitis due to nickel and chromate, *Arch. Dermatol. Syphilol.,* 73, 236, 1956.
99. **Fandejev, L. I. and Borodina, S. Z.,** Berufsdermatosen hervorgerufen durch sechswertiges Chrom bei Arbeitern der Maschinenwerke, *Excerpta Lect. Int. Symp. Dermat., Prag,* 1960, 64.
100. **Frey, E.,** Beitrag zur Bewertung der Jaegerschen Kalium-bichromatprobe auf Zement, *Dermatologica,* 105, 244, 1952.
101. **Klauder, J. V. and Combes, F. C.,** Aspects of occupational dermatoses with particular reference to treatment and chromate dermatitis, *Ind. Med. Surg.,* 24, 13, 1955.
102. **Mali, J. W. H.,** Chromium or chromate dermatitis, *Dermatologica,* 132, 175, 1976.
103. **Samitz, M. H., Gross, S., and Katz, S.,** Inactivation of chromium ion in allergic eczematous dermatitis, *J. Invest. Dermatol.,* 38, 5, 1962.
104. **Shanon, J.,** Pseudo-atopic dermatitis, Contact dermatitis due to chrome sensitivity simulating atopic dermatitis, *Dermatologica,* 131, 176, 1965.
105. **Bockendahl, H.,** Chromnachweis und Chromgehalt gefärbter Kleiderstoffe, *Dermatol. Wochenschr.,* 130, 987, 1954.
106. **Cohen, H. A.,** Carrier specificity of tuberculin-type reaction to trivalent chromium, *Arch. Dermatol.,* 93, 34, 1966.
107. **Cohen, H. A.,** Tuberculin-type reaction to heparin-chromium complex, *Arch. Dermatol.,* 94, 409, 1966.
108. **Fregert, S. and Rorsman, H.,** Allergy to trivalent chromium, *Arch. Dermatol.,* 90, 4, 1964.

109. **Fregert, S. and Rorsman, H.**, Patch test reactions to basic chromium (III) sulfate, *Arch. Dermatol.*, 91, 233, 1965.
110. **Fregert, S. and Rorsman, H.**, Allergic reactions to trivalent chromium compounds, *Arch. Dermatol.*, 93, 711, 1966.
111. **Morris, G. E.**, The manufacture of leather: its chemical and dermatological aspects, *Annu. Meet. Am. Acad. Derm. Syph. (Chicago)*, December 1956.
112. **Morris, G. E.**, "Chrome" dermatitis: a study of the chemistry of shoe leather with particular reference to basic chromic sulfate, *Arch. Dermatol.*, 78, 612, 1958.
113. **Morris, G. E.**, Sweat band dermatitis, JAMA, 169, 1747, 1959.
114. **Morris, G. E.**, Basisches Chrom-III-Sulfat. Ergebnisse der Epikutantestung eines Chrom-III-Salzes, *Berufs-Dermatosen*, 7, 126, 1959.
115. **Skog, E.**, Positive patch test to trivalent chromium, *Acta Derm.-Venereol.*, 35, 393, 1955.
116. **Negulescu, V. N.**, Sur la sensibilisation expérimentale au chrome, *XIII. Congressus Internationalis Dermatologiae, München, 1967*, Vol. 1, Springer-Verlag, Berlin, 259, 1968.
117. **Spier, H. W., Natzel, R., and Pascher, G.**, Das Chromatekzem, mit besonderer Berücksichtigung der ätiologischen Bedeutung des Spurenchromates, *Archgewerbehyg. Gewerbepathol.*, 14, 373, 1956.
118. **Carrie, C.**, *Praktischer Leitfaden der beruflichen Hautkrankheiten*, G. Thieme, Stuttgart, 1951.
119. **Hansen, K.**, *Allergie*, 3rd ed., G. Thieme, Stuttgart, 1957.
120. **van Neer, F. C.**, Reacties op intracutane injecties van drie en zeswaardige chroomverbindingen bij gesensibiliseerde mensen, varkens en cavias, *Ned. Tijdschr. Geneeskd.*, 109, 1684, 1965.
121. **Valér, M. and Rácz, I.**, Investigations concerning the sensitizing effect of trivalent chromium salts I. Correlation between concentration and skin penetration of trivalent chromium, *Berufs-Dermatosen*, 19, 302, 1971.
122. **Polak, L.**, Immunological aspects of contact sensitivity, *Monogr. Allergy*, 15, 1980.
123. **Bloch, B. and Steiner-Wourlisch, A.**, Die Sensibilisierung des Meerschweinchens gegen Primeln, *Arch. Dermatol. Syphilol.*, 162, 349, 1930.
124. **Jadassohn, W.**, Sensibilisierung der Haut des Meerschweinchens auf Phenylhydrazin, *Klin. Wochenschr.*, 12, 551, 1930.
125. **Wedroff, N. S. and Dolgoff, A. P.**, Ueber die spezifische Sensibilität der Haut einfachen chemischen Stoffen gegenüber, *Arch. Dermatol. Syphilol.*, 171, 647, 1935.
126. **Asherson, G. L. and Ptak, W.**, Contact and delayed hypersensitivity in the mouse. I. Active sensitization and passive transfer, *Immunology*, 15, 405, 1968.
127. **Polak, L.**, Self molecules in induction of hypersensitivity and tolerance in DNCB-contact sensitivity in guinea pig, *Ann. N.Y. Acad. Sci.*, in press.
128. **Silberberg, I., Baer, R. L. and Rosenthal, S.**, The role of Langerhans cells in allergic contact hypersensitivity. A review of findings in man and guinea pigs, *J. Invest. Dermatol.*, 66, 210, 1976.
128a. **Polak, L.**, unpublished results.
129. **Gell, P. G. H. and Benacerraf, B.**, Studies on hypersensitivity. II. Delayed hypersensitivity to denatured proteins in guinea pigs, *Immunology*, 2, 64, 1959.
130. **Salvin, S. B. and Smith, R. F.**, The specificity of allergic reactions. III. Contact hypersensitivity, *J. Exp. Med.*, 114, 185, 1961.
131. **Silberberg-Sinakin, I. and Baer, R. L.**, Langerhans cells. A review of their nature with emphasis on their immunologic functions, *Prog. Allergy*, 24, 268, 1978.
132. **Frey, J. R. and Wenk, P.**, Experimental studies on the pathogenesis of contact eczema in the guinea-pig, *Int. Arch. Allergy Appl. Immunol.*, 11, 81, 1957.
133. **Turk, J. L. and Stone, S. H.**, Implications of the cellular changes in lymph nodes during the development and inhibition of delayed type hypersensitivity, in *Cell-Bound Antibodies*, Amos, B. and Koprowski, H., Eds., Wistar Institute Press, Philadelphia, 1963, 51.
134. **Polak, L.**, In vitro DNA synthesis in lymphocytes from DNCB contact-sensitive guinea pigs, *Exp. Cell. Biol.*, 48, 356, 1980.
135. **Polak, L. and Scheper, R. J.**, In vitro DNA synthesis in lymphocytes from guinea pigs epicutaneously sensitized with DNCB, *J. Invest. Dermatol.*, 76, 133, 1981.
136. **Frey, J. R. and Wenk, P.**, Ueber die Funktion der regionalen Lymphknoten bei der Entstehung des Dinitrochlorbenzol-Kontaktekzems am Meerschweinchen, *Dermatologica*, 116, 243, 1958.
137. **Dumonde, D. C., Wolstencroft, R. A., Panayi, G. S., Matthew, M., Morley, J., and Howson, W. T.**, "Lymphokines": Non-antibody mediators of cellular immunity generated by lymphocyte activation, *Nature (London)*, 224, 38, 1969.
138. **Polak, L. and Rufli, T.**, Vasoactive mediators in contact sensitivity, in *New Trends in Allergy*, Ring, J. and Burg, G., Eds., Springer-Verlag, Berlin, 1981, 187.
139. **Polak, L.**, Studies on the role of suppressor cells in specific unresponsiveness to DNCB, *Immunology*, 31, 425, 1976.

140. **Asherson, G. L., Zembala, M., Thomas, W. R., and Perera, M.A.C.C.**, Suppressor cells and the handling of antigen, *Immunol. Rev.*, 50, 3, 1980.
141. **Polak, L. and Rinck, C.**, Effect of the elimination of suppressor cells on the development of DNCB contact sensitivity in guinea pigs, *Immunology*, 33, 305, 1977.
142. **Katz, S. I., Parker, D., Sommer, G., and Turk, J. L.**, Suppressor cells in normal immunisation as a basic homeostatic phenomenon, *Nature (London)*, 248, 612, 1974.
143. **Polak, L.**, The transfer of tolerance to DNCB-contact sensitivity in guinea pigs by parabiosis, *J. Immunol.*, 114, 988, 1975.
144. **Polak, L. and Turk, J. L.**, Reversal of immunological tolerance by Cyclophosphamide through the inhibition of suppressor cell activity, *Nature (London)*, 249, 654, 1974.
145. **Magnusson, B. and Kligman, A. M.**, *Allergic contact dermatitis in the guinea pig; Identification of Contact Allergens*, Charles C Thomas, Springfield, Ill., 1970.
146. **Darabos, L., Szabó, Z., and Megyeri, J.**, Beiträge zur Bichromatallergie. IV. Tierexperimentelle Prüfungen unter besonderer Berücksichtigung der Arbeitsverhältnisse in Schieferfabriken, *Hautarzt*, 14, 364, 1963.
147. **Heise, H., Mattheus, A., and Flegel, H.**, Sensibilisierungsversuche gegen Chromverbindungen unter Mitwirkung von DMSO, *Allerg. Asthma*, 15, 151, 1969.
148. **Levin, H. M., Brunner, M. J., and Rattner, H.**, Lithographer's dermatitis, *JAMA*, 169, 566, 1959.
149. **Mayer, R. L. and Jaconia, D.**, Zur Frage der Chromüberempfindlichkeit, *Allerg. Asthma*, 4, 275, 1958.
150. **Nilzen, A. and Wikström, K.**, The influence of lauryl sulphate on the sensitization of guinea pigs to chrome and nickel, *Acta Derm.-Venereol.*, 35, 292, 1955.
151. **Scheidegger, J. P., Schwarz-Speck, M., Schwarz, K., and Storck, H.**, Experimentelles Ekzem auf Kaliumbichromat im Tierversuch, *Dermatologica*, 135, 382, 1967.
152. **Vallecchi, C. and Giannotti, B.**, Die experimentelle Chromdermatitis beim Meerschweinchen, *Z. Haut-Geschlechtskr.*, 124, 11, 1968.
153. **Wikström, K.**, Epidermal treatment of guinea pigs with potassium bichromate, *Acta Derm.-Venereol.*, 42 (Suppl. 49), 1, 1962.
154. **Schwarz-Speck, M.**, Experimentelle Sensibilisierung mit drei- und sechswertigem Chrom, XIII. Congressus Internationalis Dermatologiae, München, 1967, Springer-Verlag, Berlin, 1968.
155. **Samitz, M. H. and Pomerantz, H.**, Studies of the effects on the skin of nickel and chromium salts, *AMA Arch. Ind. Health*, 18, 473, 1958.
156. **Cohen, H. A.**, On the pathogenesis of contact dermatitis, *Isr. J. Med. Sci.*, 2, 37, 1966.
157. **Gross, P. R., Katz, S. A., and Samitz, M. H.**, Sensitization of guinea pigs to chromium salts, *J. Invest. Derm.*, 50, 424, 1968.
158. **Hunziker, N.**, A propos de l'hypersensibilité au bichromate de potassium chez le cobaye, *Dermatologica*, 121, 93, 1960.
159. **van Neer, F. C. J.**, Sensitization of guinea pigs to chromium compounds, *Nature (London)*, 198, 1013, 1963.
160. **Polak, L. and Turk, J. L.**, Studies on the effect of systematic administration of sensitizers in guinea-pigs with contact sensitivity to inorganic metal compounds. I. The induction of immunological unresponsiveness in already sensitized animals, *Clin. Exp. Immunol.*, 3, 245, 1968.
161. **Skog, E. and Wahlberg, J. E.**, Sensitization and testing of guinea pigs with potassium bichromate, *Acta Derm.-Venereol.*, 50, 103, 1970.
162. **Hicks, R., Hewitt, P. J., and Lam, H. F.**, An investigation of the experimental induction of hypersensitivity in the guinea pig by material containing chromium, nickel and cobalt from arc welding fumes, *Int. Arch. Allergy Appl. Immunol.*, 59, 265, 1979.
163. **Maguire, H. C. and Chase, M. W.**, Exaggerated delayed-type hypersensitivity to simple chemical allergens in the guinea pig, *J. Invest. Dermatol.*, 49, 460, 1967.
164. **Darabos, L.**, Beiträge zur Bichromat-Allergie. II. Die Modifizierung der Bichromatprobe, *Hautarzt*, 11, 408, 1960.
165. **Gaul, L. E.**, Comparison of patch and contact test response in chromate sensitivity, *Ann. Allergy*, 13, 243, 1955.
166. **Meneghini, C. L. and Giannotti, F.**, Studi sull'allergie al cromo in soggetti con eczema da cemento, *G. Ital. Dermatol.*, 97, 225, 1955.
167. **Frey, J. R., de Weck, A. L., and Geleick, H.**, Sensitization, immunological tolerance and desensitization of guinea pigs to neoarsphenamine. II. Influence of various factors on sensitization to NEO, *Int. Arch. Allergy*, 30, 385, 1966.
168. **Skog, E. and Wahlberg, J. E.**, Passive transfer of chromium allergy in guinea-pigs, *Acta Derm.-Venereol.*, 50, 189, 1970.

169. **Shelley, W. B. and Juhlin, L.**, Langerhans cells form a reticuloepithelial trap for external contact antigens, *Nature (London)* 261, 46, 1976.
170. **van Kooten, W. J., Mali, J. W. H., de Goeij, J. J. M., and Houtman, J. P. W.**, Determination of the chromium content in human skin by means of neutron activation analysis, Symp. Nucl. Activ. Tech. Life Sci. Amsterdam, May 1967.
171. **Sertoli, A. and Panconesi, E.**, Investigations on the pathogenesis of chromium salts eczema, *XIII. Congressus Internationalis Dermatologiae, München, 1967*, Springer-Verlag, Berlin, 261, 1968.
172. **Czernilewski, A., Brykolski, D., and Depczyk, D.**, Experimental investigations on penetration of radioactive chromium (^{51}Cr) through the skin, *Dermatologica*, 131, 334, 1965.
173. **Mali, J. W. H., van Kooten, W. J., Spruit, D., and van Neer, F. C. J.**, Quantitative aspects of chromium sensitization, *Acta Derm.-Venereol.*, 44, 44, 1964.
174. **Schwarz, E.**, Experimentelle Untersuchungen zum Chromatekzem, *Arch. Klin. Exp. Dermatol.*, 213, 493, 1961.
175. **Schwarz, E. and Spier, H. W.**, Die percutane Resorption von 3- und 6-wertigem Chrom (Cr^{51}). Zur Pathogenese des Kontaktekzems, *Arch. Klin. Exp. Derm.*, 210, 20, 1960.
176. **Wahlberg, J. E. and Skog, E.**, The percutaneous absorption of sodium chromate (^{51}Cr) in the guinea pig, *Acta Derm.-Venereol.*, 43, 102, 1963.
177. **Wahlberg, J. E. and Skog, E.**, Percutaneous absorption of tri- and hexavalent chromium, *Arch. Dermatol.*, 92, 315, 1965.
177a. **Blank, St.**, personal communication.
178. **Mali, J. W. H., Spruit, D., and van Neer, F. C. J.**, Bemerkungen zum Chromatekzem, *Allerg. Asthma*, 9, 294, 1963.
179. **Frey, J. R., de Weck, A. L., Geleick, H., and Polak, L.**, The induction of immunological tolerance during the primary response, *Int. Arch. Allergy Appl. Immunol.*, 42, 278, 1972.
180. **Wahlberg, J. E.**, Disappearance measurements: A method for studying percutaneous absorption of isotope-labelled compounds emitting γ-rays, *Acta Derm.-Venereol.*, 45, 397, 1965.
181. **Skog, E. and Wahlberg, J. E.**, A comparative investigation of the percutaneous absorption of metal compounds in the guinea pig by means of the radioactive isotopes: Cr^{51}, Co^{58}, Zn^{65}, Ag^{110m}, Cd^{115m}, Hg^{203}, *J. Invest. Dermatol.*, 43, 187, 1964.
182. **Wahlberg, J. E.**, Percutaneous absorption of sodium chromate (^{51}Cr), cobaltous (^{58}Co) and mercuric (^{203}Hg) chlorides through excised human and guinea pig skin, *Acta Derm.-Venereol.*, 45, 415, 1965.
183. **Wahlberg, J. E.**, Percutaneous absorption from chromium (^{51}Cr) solutions of different pH, 1.4-12.8. An experimental study in the guinea pig, *Dermatologica*, 137, 17, 1968.
184. **Wahlberg, J. E.**, Skin clearance of iontophoretically administered chromium (Cr) and sodium (Na) ions in the guinea pig, *Acta Derm.-Venereol.*, 50, 255, 1970.
185. **Wahlberg, J. E.**, Percutaneous absorption of trivalent and hexavalent chromium (^{51}Cr) through excised human and guinea pig skin, *Dermatologica*, 141, 288, 1970.
186. **van Kooten, W. J. and van Neer, F. C. J.**, Resorption of chromium compounds in guinea-pigs after intradermal injection and iontophoresis, *Dermatologica*, 132, 183, 1966.
187. **Heise, H. and Mathäus, A.**, Immunofluoreszenz-Untersuchungen bei verschiedenen Dermatosen. IV. Antigennachweis am positiven Kaliumbichromattest mit der direkten Immunofluoreszenz-Methode, *Arch. Klin. Exp. Dermatol.*, 231, 239, 1968.
188. **Pedersen, N. B., Bertilsson, G., Fregert, S., Lidén, K., and Rorsman, H.**, Disappearance of chromium injected intracutaneously, *Int. Arch. Allergy Appl. Immunol.*, 36, 82, 1969.
189. **Pedersen, N. B., Fregert, S., Naversten, Y., and Rorsman, H.**, Patch testing and absorption of chromium, *Acta Derm.-Venereol.*, 50, 431, 1970.
190. **Ziegler, G., Lüthy, H., and Gähwiler, B.**, Die Verweildauer radioaktiver Metallionen auf der Haut, *Dermatologica*, 137, 259, 1968.
191. **Christie, O. H. J., Dinh-Nguyen, N., Vincent, J., Hellgren, L., and Pimlott, W.**, Spark source mass spectrographic study of metal allergenic substances on the skin, *J. Invest. Dermatol.*, 67, 587, 1976.
192. **Thormann, J., Jespersen, N. B., and Joensen, H. D.**, Persistence of contact allergy to chromium, *Contact Dermatitis*, 5, 261, 1979.
193. **Pedersen, N. B. and Naversten, N.**, Disappearance of chromium (III) trichloride injected intracutaneously, *Acta Derm.-Venereol.*, 53, 127, 1973.
194. **Mali, J. W. H., van Kooten, W. J., and van Neer, F. C. J.**, Some aspects of the behavior of chromium compounds in the skin, *J. Invest. Dermatol.*, 41, 111, 1963.
195. **Spruit, D. and van Neer, F. C.**, Penetration rate of Cr (III) and Cr (VI), *Dermatologica*, 132, 179, 1966.
196. **Samitz, M. H. and Shrager, J.**, Patch test reaction to hexavalent and trivalent chromium compounds, *Arch. Dermatol.*, 94, 304, 1966.

197. **Samitz, M. H., Shrager, J. D., and Katz, S.**, Chemical reactions between chromium and skin: studies of hypersensitivity to chromium ion., Symp. Derm. "De Structura et Functione Stratorum Epidermidis S. D. Barrierae," Brno, 1964.
198. **Samitz, M. H., Katz, S., and Shrager, J. D.**, Studies of the diffusion of chromium compounds through skin, *J. Invest. Dermatol.*, 48, 514, 1967.
199. **Forslind, B. and Wahlberg, E.**, The morphology of chromium allergic skin reactions at electron microscopic resolution: studies in man and guinea pigs, *Acta Derm.-Venereol.*, 58, Suppl. 79, 43, 1978.
200. **Gray, S. J. and Stirling, K.**, The tagging of red cells and plasma proteins with radioactive chromium, *J. Clin. Invest.*, 29, 1604, 1950.
201. **Visek, W. J., Whitney, I. B., Kuhn, U. S. G., and Comar, C. L.**, Metabolism of Cr by animals as influenced by chemical state, *Proc. Soc. Exp. Biol. Med.*, 84, 610, 1953.
202. **Baetjer, A. M., Damron, C., and Budacz, V.**, The distribution and retention of chromium in men and animals, *AMA Arch. Ind. Health*, 20, 136, 1959.
203. **Hueper, W. C. and Payne, W. W.**, Experimental cancers in rats produced by chromium compounds and their significance to industry and public health, *Ind. Hyg. J.*, 20, 274, 1959.
204. **Jandl, J. H. and Simmons, R. L.**, The agglutination and sensitization of red cells by metallic cations: interaction between multivalent metals and the red cell membrane, *Br. J. Haematol.*, 3, 19, 1957.
205. **Cohen, Y., Wepierre, J., and Ponty, D.**, Fixation du chrome-51 sur les fractions protéiques du sérum sanguin, *Radioact. Isot. Klin. Forsch.*, 6, 274, 1965.
206. **Hopkins, L. L., Jr. and Schwarz, K.**, Chromium (III) binding to serum proteins, specifically siderophillin, *Biochim. Biophys. Acta*, 90, 484, 1964.
207. **Bauer, J. A.**, Genetics of skin transplantation and an estimate of the number of histocompatibility genes in inbred guinea pigs, *Ann. N. Y. Acad. Sci.*, 87, 78, 1960.
208. **Samitz, M. H. and Katz, S.**, Preliminary studies on the reduction and binding of chromium with skin, *Arch. Dermatol.*, 88, 816, 1963.
209. **Samitz, M. H., Katz, S. A., Schreiner, D. M., and Gross, P. R.**, Chromium-protein interactions, *Acta Derm.-Venereol.*, 49, 142, 1969.
210. **Anderson, F. E.**, Biochemical experiments on the binding of chrome to skin, *Br. J. Dermatol.*, 72, 149, 1960.
211. **Grogan, C. H., Cahnmann, H. J., and Lethco, E.**, Microdetermination of chromium in small samples of various biological media, *Anal. Chem.*, 27, 983, 1955.
212. **Urone, P. F. and Anders, H. K.**, Determination of small amounts of chromium in human blood, tissues and urine. (Colorimetric method), *Anal. Chem.*, 22, 1317, 1950.
213. **Magnus, I. A.**, The conjugation of nickel, cobalt, hexavalent chromium and eosin with protein as shown by paper electrophoresis, *Acta Derm.-Venereol.*, 38, 220, 1958.
214. **Katz, S. A., Scheiner, D. M., Klein, A. W., and Samitz, M. H.**, Chromium complexes with proteins and mucopolysaccharides and their relationship to chromium allergy in sensitized guinea pigs, *Environ. Res.*, 7, 212, 1974.
215. **Herrmann, H. and Speck, L. B.**, Interaction of chromate with nucleic acids in tissues, *Science*, 119, 221, 1954.
216. **Samitz, M. H. and Katz, S.**, A study of the chemical reactions between chromium and skin, *J. Invest. Dermatol.*, 43, 35, 1964.
216a. **Kallos, P.**, personal communication.
217. **Cohen, H. A.**, Experimental production of circulating antibodies to chromium, *J. Invest. Dermatol.*, 38, 13, 1962.
218. **Heise, H. and Mattheus, A.**, Immunofluoreszenz-Untersuchungen bei verschiedenen Dermatosen. I. Herstellung fluorochrommarkierter Antiseren, *Arch. Klin. Exp. Dermatol.*, 226, 420, 1966.
219. **Heise, H. and Mattheus, A.**, Immunofluoreszenz-Untersuchungen bei verschiedenen Dermatosen. II. Untersuchungen an Probeexcisionen aus positiven Kalium-bichromat-Epicutantesten mit markiertem anti-gamma-Globulin, *Arch. Klin. Exp. Dermatol.*, 228, 14, 1967.
220. **Winter, V. and Freund, D.**, Untersuchungen des experimentellen Chromekzems beim Meerschweinchen mit Hilfe fluoreszierender Antikörper, *Berufs-Dermatosen*, 15, 176, 1967.
221. **Penders, A. J. M. and Grosfeld, J. C. M.**, Serological observations on patients with chromium-eczema and chromium-sensibilized guinea-pigs, *Dermatologica*, 132, 177, 1966.
222. **Polak, L., Frey, J. R., and Turk, J. L.**, Studies on the effect of systemic administration of sensitizers in guinea-pigs with contact sensitivity to inorganic metal compounds. V. Studies on the mechanism of the flare-up reaction, *Clin. Exp. Immunol.*, 7, 739, 1970.
223. **Grogan, C. H. and Oppenheimer, H.**, Experimental studies in metal cancerogenesis. III. Behavior of chromium compounds in the physiological pH range, *J. Am. Chem. Soc.*, 77, 152, 1955.

224. **Holz, H., Mappes, R., and Weidmann, G.**, Chromatallergie bei Bohrölekzem, *Berufs-Dermatosen*, 9, 113, 1961.
225. **Schneeberger, H. W. and Forch, G.**, Tierexperimentelle Chromallergie. Einfluss der Valenzen und der chemischen Umgebung der Chromionen, *Arch. Dermatol., Forsch.*, 249, 71, 1974.
226. **Shmunes, E., Katz, S. A., and Samitz, M. H.**, Chromium-amino acid conjugates as elicitors in chromium-sensitized guinea pigs, *J. Invest. Dermatol.*, 60, 139, 1973.
227. **Polak, L.**, Mechanism of reappearance of contact sensitivity to DNCB in desensitized guinea pigs, *J. Invest. Dermatol.*, 66, 38, 1976.
228. **Thulin, H. and Zachariae, H.**, The leucocyte migration test in chromium hypersensitivity, *J. Invest. Dermatol.*, 58, 55, 1972.
229. **Thomas, D. W. and Shevach, E. M.**, Nature of the antigenic complex recognized by T lymphocytes. I. Analysis with an in vitro primary response to soluble protein antigens, *J. Exp. Med.*, 144, 1263, 1976.
230. **Miller, A. E. and Levis, W. R.**, Studies on the contact sensitization of man with simple chemicals. I. Specific lymphocyte transformation in response to DNCB sensitization, *J. Invest. Dermatol.*, 61, 261, 1973.
231. **Aspergen, N. and Rorsman, H.**, Short-term culture of leucocytes in nickel hypersensitivity, *Acta Derm.-Venereol.*, 42, 412, 1962.
232. **Bizzozero, E. and Depaoli, M.**, Ueber die Histopathogenese der allergischen Hautreaktionen äusseren Ursprungs, *Hautarzt*, 7, 487, 1956.
233. **Everett, E. T., Livingood, C. S., Pomerat, C. M., and Hu, F.**, Tissue culture studies on human skin. II. Comparative effects of certain specific contact allergens on sensitized and non sensitized human skin, *J. Invest. Dermatol.*, 18, 193, 1952.
234. **Grosfeld, J. C. M., Penders, A. J. M., de Grood, R., and Verwilchen, L.**, In vitro investigations of chromium- and nickel-hypersensitivity with culture of skin and peripheral lymphocytes, *Dermatologica*, 132, 189, 1966.
235. **Jung, E. G.**, Die Bedeutung der Lymphoblastentransformation beim allergischen Kontaktekzem, *Ther. Umsch.*, 26, 94, 1969.
236. **Schöpf, E., Schulz, K. H., und Isensee, I.**, Untersuchungen über den Lymphocytentransformationstest bei Quecksilberallergie. Unspezifische Transformation durch Hg Verbindungen, *Arch. Klin. Exp. Dermatol.*, 234, 420, 1969.
237. **Pappas, A., Orfanos, C. D., and Bertram, R.**, Non-specific lymphocyte transformation in vitro by nickel acetate, *J. Invest. Dermatol.*, 55, 198, 1970.
238. **Bennet, B. and Bloom, B. R.**, Reactions in vivo and in vitro produced by a soluble substance associated with delayed type hypersensitivity, *Proc. Natl. Acad. Sci. U.S.A.* 59, 756, 1968.
239. **Pick, E., Brostoff, J., Krejčí, J., and Turk, J. L.**, Interaction between "sensitized lymphocytes" and antigen in vitro. II. Mitogen-induced release of skin reactive and macrophage migration inhibitory factors, *Cell. Immunol.* 1, 92, 1970.
240. **Pick, E., Krejčí, J., Čech, K., and Turk, J. L.**, Interaction between sensitized lymphocytes and antigen in vitro. I. The release of skin reactive factor, *Immunology*, 17, 741, 1969.
241. **Pick, E., Krejčí, J., and Turk, J. L.**, Release of skin reactive factor from guinea-pig lymphocytes by mitogens, *Nature (London)*, 225, 236, 1970.
242. **Maillard, J. L., Pick, E., and Turk, J. L.**, Interaction between "sensitized lymphocytes" and antigen in vitro. III. Vascular permeability, induced by skin reactive factor (SRF), *Int. Arch. Allergy Appl. Immunol.*, 42, 50, 1972.
243. **Bloom, B. R. and Bennett, B.**, The assay of inhibition of macrophage migration and the production of migration inhibitory factor (MIF) and skin reactive factor (SRF) in the guinea pig, in *In Vitro Methods in Cell-Mediated Immunity*, Bloom, B. R. and Glade, P. R., Eds., Academic Press, New York, 1971, 235.
244. **Pick, E., Krejčí, J., and Turk, J. L.**, In vitro production and assessment of activity of skin reactive factors, in *In Vitro Methods in Cell-Mediated Immunity*, Bloom, B. R. and Glade, P. R., Eds., Academic Press, New York, 1971, 515.
245. **David, J. R., Askari, A. S., Lawrence, H. S., and Thomas, C.**, Delayed hypersensitivity in vitro. I. Specificity of inhibition of cell migration by antigens, *J. Immunol.*, 93, 264, 1964.
246. **Bloom, B. R. and Bennett, B.**, Delayed hypersensitivity *in vitro*: the mechanism of inhibition by antigen of cell migration, *Fed. Proc. Fed. Am. Soc. Exp. Biol.*, 25, 355, 1966.
246a. **Polak, L.**, unpublished results.
247. **Cohen, H. A.**, Tuberculin-type sensitivity to trivalent chromium, *Isr. J. Med. Sci.*, 1, 79, 1965.
248. **Zelger, J. and Michelmayr, G.**, Lymphokine bei Chromatallergie, *Allergol. Immunopathol.*, 5, 358, 1977.

249. **Bloch, B. and Steiner-Wourlisch, A.**, Die willkürliche Erzeugung der Primelüberempfindlichkeit beim Menschen und ihre Bedeutung für das Idiosynkrasieproblem, *Arch. Dermatol., Syph.,* 152, 283, 1926.
250. **Parker, D., Sommer, G., and Turk, J. L.**, Variation in guinea pig responsiveness, *Cell. Immunol.,* 18, 233, 1975.
251. **Polak, L. and Turk, J. L.**, Genetic background of certain immunological phenomena with particular reference to the skin, *J. Invest. Dermatol.,* 52, 219, 1969.
252. **Chase, M. W.**, Inheritance in guinea pigs of the susceptibility to skin sensitization with simple chemical compounds, *J. Exp. Med.,* 73, 711, 1941.
253. **Schwartz, B. D., Kask, A. M., Paul, W. E., Geczy, A. F., and Shevach, E. M.**, The guinea pig I region. I. A structural and genetic analysis, *J. Exp. Med.,* 146, 547, 1977.
254. **Shevach, E. M., Lundquist, M., Geczy, A. F., and Schwartz, B. D.**, The guinea pig I region. II. Functional analysis, *J. Exp. Med.,* 146, 561, 1977.
255. **Geczy, A. F. and de Weck, A. L.**, Molecular basis of T cell dependent genetic control of the immune response in the guinea pig, *Progr. Allergy,* 22, 147, 1977.
255a. **Siegenthaler, U. and Polak, L.**, unpublished results.
256. **Ben Efraim, S., Fuchs, S., and Sela, M.**, Differences in immune response to synthetic antigens in two inbred strains of guinea pigs, *Immunology,* 12, 573, 1967.
257. **Ben Efraim, S. and Maurer, P. H.**, Immune response to polypeptides (poly-L-amino acids) in inbred guinea pigs, *J. Immunol.,* 97, 577, 1966.
258. **Levine, B. B., Ojeda, A., and Benacerraf, B.**, Studies on artificial antigens. III. The genetic control of the immune response to hapten poly-L-lysine conjugates in guinea pigs, *J. Exp. Med.,* 118, 953, 1963.
259. **Stingl, G., Katz, S. I., Clement, L., Green, I., and Shevach, E. M.**, Immunologic functions of Ia-bearing epidermal Langerhans cells, *J. Immunol.,* 121, 2005, 1978.
260. **Polak, L., Barnes, J. M., and Turk, J. L.**, The genetic control of contact sensitization to inorganic metal compounds in guinea-pigs, *Immunology,* 14, 707, 1968.
261. **Baumgarten, A. and Geczy, A. F.**, Induction of delayed hypersensitivity by dinitrophenylated lymphocytes, *Immunology,* 19, 205, 1970.
262. **Levey, R. H. and Medawar, P. B.**, Nature and mode of action of anti-lymphocyte serum, *Proc. Natl. Acad. Sci. U.S.A.* 56, 1130, 1966.
263. **Burnet, F. M. and Fenner, F.**, *The Production of Antibodies,* McMillan, Melbourne, 1949.
264. **Sulzberger, M. B. and Mayer, R. L.**, Sensitizations. Regional, seasonal, dietary and other influences accounting for variations and fluctuations, *Arch. Dermatol. Syphilol.,* 24, 537, 1931.
265. **Sulzberger, M. B.**, Hypersensitiveness to arsphenamine in guinea pigs. I. Experiments in prevention and in desensitization, *Arch. Dermatol., Syphilol.,* 20, 669, 1929.
266. **Chase, M. W.**, Inhibition of experimental drug allergy by prior feeding of the sensitizing agent, *Proc. Soc. Exp. Biol. Med.,* 61, 257, 1946.
267. **Medawar, P. B.**, *Theories of Immunological Tolerance,* Wolstenholme, G. E. W. and O'Connor, M., Eds., Ciba Foundation Symp. Cell. Aspects Immun. J. & A. Churchill, 1960, 134.
268. **Medawar, P. B.**, Tolerance reconsidered—a critical survey, *Transplant. Proc.,* 5, 7, 1973.
269. **Frey, J. R., de Weck, A. L., and Geleick, H.**, Inhibition of the contact reaction to dinitrochlorobenzene by intravenous injection of dinitrobenzene sulfonate in guinea pigs sensitized to dinitrochlorobenzene, *J. Invest. Dermatol.,* 42, 189, 1964.
270. **de Weck, A. L. and Frey, J. R.**, *Immunotolerance to Simple Chemicals,* S. Karger, Basel, 1966.
271. **Polak, L.**, Recent trends in the immunology of contact sensitivity. II, *Contact Dermatitis,* 4, 256, 1978.
272. **Asherson, G. L., Zembala, M., Thomas, W. R., and Perera, M. A. C. C.**, Suppressor cells and the handling of antigen, *Immunol. Rev.,* 50, 3, 1980.
273. **Claman, H. N., Miller, S. D., Sy, M. S., and Moorhead, J. W.**, Suppressive mechanisms involving sensitization and tolerance in contact sensitivity, *Immunol. Rev.,* 50, 105, 1980.
274. **Polak, L. and Geleick, H.**, Differing mechanisms of tolerance and desensitization to DNCB in guinea pigs, *Eur. J. Immunol.,* 5, 94, 1975.
275. **Poulter, L. W. and Turk, J. L.**, Changes in macrophages in vivo induced by densensitization, *Cell. Immunol.,* 23, 171, 1976.
276. **Dwyer, J. M. and Kantor, F. S.**, In vivo suppression of delayed hypersensitivity: prolongation of desensitization in guinea pigs, *J. Exp. Med.,* 142, 588, 1975.
277. **Liew, F. Y.**, Densitization of delayed type hypersensitivity by antigen and specific antibody, *Cell. Immunol.,* 19, 129, 1975.
278. **Katayama, I. and Nishioka, K.**, Suppression of contact sensitivity by IgGl antihapten antibody in contact-sensitized guinea pigs, *J. Invest. Dermatol.,* 76, 197, 1981.

279. **de Weck, A. L., Frey, J. R., and Geleick, H.,** Specific inhibition of contact dermatitis to dinitrochlorobenzene in guinea pigs by injection of haptens and protein conjugates, *Int. Arch. Allergy Appl. Immunol.,* 24, 63, 1964.
280. **Phanuphak, P., Moorhead, J. W., and Claman, H. N.,** Tolerance and contact sensitivity to DNCB in mice. IV. Densensitization as a manifestation of increased proliferation of sensitized cells, *J. Immunol.,* 114, 1147, 1975.
281. **Parker, D., Turk, J. L., and Scheper, R. J.,** Central and peripheral action of suppressor cells in contact sensitivity in the guinea pig, *Immunology,* 30, 593, 1976.
282. **Asherson, G. L. and Zembala, M.,** Suppression of contact sensitivity by T cells in the mouse. I. Demonstration that suppressor cells act on the effector stage of contact sensitivity; and their induction following in vitro exposure to antigen, *Proc. R. Soc. Med.,* 187, 320, 1974.
283. **Polak, L. and Rinck, C.,** Mechanism of desensitization in DNCB-contact sensitive guinea pigs, *J. Invest. Dermatol.,* 70, 98, 1978.
284. **Akimov, V. G.,** On specific inhibition of contact allergic reactions experimentally, *Vestn. Dermatol. Venereol.,* 12, 20, 1973.
285. **Lowney, E. D.,** Immunologic unresponsiveness to a contact sensitizer in man, *J. Invest. Dermatol.,* 51, 411, 1968.
286. **Skog, E.,** Theoretical and practical aspects of hyposensitization in cutaneous allergies, Trans. 12th Int. Congr. Dermatol., Washington, 1962, 1112.
287. **Hunziker, N.,** *Experimental Studies on Guinea Pig Eczema. Their Significance in Human Eczema,* Springer, Berlin, 1969.
288. **Frey, J. R., de Weck, A. L., and Geleick, H.,** Studies on the induction of immunological tolerance by antigen in guinea-pigs already sensitized to dinitrochlorobenzene, *Clin. Exp. Immunol.,* 8, 131, 1971.
289. **Gell, P. G. H. and Benacerraf, B.,** Studies on hypersensitivity. VI. The relationship between contact and delayed sensitivity: a study on the specificity of cellular immune reactions, *J. Exp. Med.,* 113, 571, 1961.
290. **Frey, J. R., de Weck, A. L., and Geleick, H.,** Sensitization, immunological tolerance and desensitization of guinea pigs to neoarsphenamine. IV. Desensitization to NEO, *Int. Arch. Allergy Appl. Immunol.,* 30, 521, 1966.
291. **Fruchard, J. and Fruchard, J.,** Nouvel essai de désensibilisation au bichromate de potassium dans un eczéma du ciment, *Bull. Soc. Fr. Dermatol. Syphiligr.,* 64, 776, 1957.
292. **Sourreil, P., Fruchard, J., and Fruchard, J.,** Désensibilisation dans un eczéma du ciment, *Bull. Soc. Fr. Dermatol. Syphiligr.,* 71, 751, 1964.
293. **Sourreil, P., Fruchard, J., and Fruchard, J.,** Nouvelle désensibilisation dans l'eczéma du ciment, *Bull. Soc. Fr. Dermatol. Syphiligr.,* 71, 752, 1964.
294. **Ado, A. D. and Sonsonkin, I. E.,** Specific desensitization in contact allergy due to metal compounds, *Vestn. Dermatol. Venereol.,* 45, 45, 1971.
295. **Chase, M. W. and Battisto, J. R.,** Immunologic unresponsiveness to allergenic chemicals, in *Mechanisms of Hypersensitivity,* Shaffer, J. H., LoGrippo, G. A., and Chase, M. W., Eds., Little, Brown, Boston, 1959, 507.
296. **Burrows, D. and Calnan, C. D.,** Cement Dermatitis: II. Clinical aspects, *Trans. St. Johns Hosp. Dermatol. Soc.,* 51, 27, 1965.
297. **Turk, J. L.,** Studies on the mechanism of action of methotrexate and cyclophosphamide on contact sensitivity in the guinea pig, *Int. Arch. Allergy Appl. Immunol.,* 24, 191, 1964.
298. **Turk, J. L.,** An experimental model for the investigation of the cellular basis of desensitization in contact sensitivity, *Int. Arch. Allergy Appl. Immunol.,* 28, 105, 1965.
299. **Turk, J. L. and Willoughby, D. A.,** Central and peripheral effects of anti-lymphocyte sera, *Lancet,* I, 249, 1967.
300. **Turk, J. L., Willoughby, D. A., and Stevens, J. E.,** An analysis of the effect of some types of anti-lymphocyte sera on contact hypersensitivity and certain models of inflammation, *Immunology,* 14, 683, 1968.
301. **Claman, H. N.,** Corticosteroids and lymphoid cells, *N. Engl. J. Med.,* 287, 388, 1972.
302. **Simmons, R. L., Ozerkis, A. J., and Hoehn, R. J.,** Antiserum to lymphocytes. Interaction with chemical immunosuppressants, *Science,* 160, 1127, 1968.
303. **Polak, L. and Frey, J. R.,** Tolerance and desensitization in experimental eczema, in *Current Problems in Dermatology,* Vol. 4, Mali, J. W. H., Ed., S. Karger, Basel, 1972, 146.
304. **Frey, J. R., de Weck, A. L., Geleick, H., and Polak, L.,** Induction of tolerance during the primary response to simple chemicals, Abstr. VIIIth Symp. Coll. Int. Allergologrium Montreux, 1970.
304a. **Lefkovits, I.,** unpublished.

305. **Frey, J. R., de Weck, A. L., Geleick, H., and Polak, L.,** Immunological tolerance in contact-hypersensitivity to dinitrochlorobenzene. Dose and time dependence. Possible cellular kinetics, *Immunology*, 21, 483, 1971.
306. **Shmunes, E. and Samitz, M. H.,** Tolerance of hexavalent chromium induced by Freund's incomplete adjuvant in guinea pigs, *Acta Derm.-Venereol.*, 53, 264, 1973.
307. **Jankovic, B. D.,** Impairment of immunological reactivity in guinea pigs by prior injection of adjuvant, *Nature (London)*, 193, 789, 1962.
308. **Polak, L.,** Suppressor cells in different types of unresponsiveness to DNCB contact sensitivity in guinea pigs, *Clin. Exp. Immunol.*, 19, 543, 1975.
309. **Chase, M. W.,** Tolerance towards chemical allergens, *Colloq. Int. C.N.R.S.*, 116, 139, 1963.
310. **Polak, L., Geleick, H., and Frey, J. R.,** The cellular mechanism of tolerance and desensitization in contact hypersensitivity to DNCB in guinea pigs, in *Contact Hypersensitivity in Experimental Animals*, Vol. 8, Monogr. Allergy, S. Karger, Basel, 1974, 168.
311. **Asherson, G. L. and Zembala, M.,** Inhibitory T cells, *Curr. Top. Microbiol. Immunol.*, 72, 55, 1975.
312. **Claman, H. N., Miller, S. D., and Moorhead, J. W.,** Tolerance: two pathways of negative immunoregulation in contact sensitivity to DNFB, *Cold Spring Harbor Symp. Quant. Biol.*, 61, 1977, 105.
313. **Polak, L., Polak, A. M., and Frey, J. R.,** Increased DNA-synthesis in vitro in guinea pigs unresponsive to DNP-skin protein conjugate, *Immunology*, 27, 115, 1974.
314. **Felton, L. D., Kaufmann, G., Prescott, B., and Ottinger, B.,** Studies on the mechanisms of the immunological paralysis induced in mice by pneumococcal polysaccharides, *J. Immunol.*, 74, 17, 1955.
315. **Howard, J.,** Properties of antigen, in *Immunological Tolerance*, Landy, M. and Braun, W., Eds., Academic Press, New York, 1969, 28.
316. **Denman, A. M., Denman, E. J., and Embling, P. H.,** Changes in the life-span of circulating small lymphocytes in mice after treatment with anti-lymphocyte globulin, *Lancet*, I, 321, 1968.
317. **Mitchison, N. A.,** Mechanism of action of antilymphocyte serum, *Fed. Proc. Fed. Am. Soc. Exp. Biol.*, 29, 222, 1970.
317a. **Polak, L. and Frey, J. R.,** unpublished observation.
318. **Polak, L., Frey, J. R., and Turk, J. L.,** Antilymphocyte serum and cyclophosphamide in the induction of tolerance to skin allografts in guinea pigs, *Transplantation*, 13, 310, 1972.
319. **Baer, R. L.,** Examples of cross-sensitization in allergic eczematous dermatitis, *Arch. Dermatol., Syphilol.*, 58, 276, 1948.
320. **Horsfall, F. L., Jr.,** Formaldehyde hypersensitiveness. An experimental study, *J. Immunol.*, 27, 569, 1934.
321. **Sidi, E. and Dobkevitch-Morrill, S.,** The injection and ingestion test in cross sensitization to the para-group, *J. Invest. Dermatol.*, 16, 299, 1951.
322. **Storck, H.,** Ekzem durch Inhalation, *Schweiz. Med. Wochenschr.*, 85, 608, 1955.
323. **Arnason, B. G. and Waksman, B. H.,** The retest reaction in delayed sensitivity, *Lab. Invest.*, 12, 671, 1963.
324. **Nakagawa, S., Fukushiro, S., Gotoh, M., Kodha, M., Namba, M., and Tanioku, K.,** Studies on the retest reaction in contact sensitivity to DNCB, *Dermatologica*, 157, 13, 1978.
325. **Nakagawa, S., Fukushiro, S., and Kohda, M.,** Antigen of retest reaction in contact sensitivity to DNCB, *Kawasaki Med. J.*, 3, 135, 1977.
326. **Dolby, A. E., Williamson, J. J., and Abrams, D.,** The retest contact hypersensitivity reaction in guinea pig skin and oral mucosa, *Br. J. Exp. Pathol.*, 50, 343, 1969.
327. **Walthard, B.,** Die Erzeugung experimenteller Nickelidiosynkrasie bei Laboratoriumstieren, *Schweiz. Med. Wochenschr.*, 7, 603, 1926.
328. **Frei, W.,** Ueber willkürliche Sensibilisierung gegen chemisch definierte Substanzen. II. Untersuchungen mit Neo-Salvarsan am Tier, *Klin. Wochenschr.*, 7, 1026, 1928.
329. **Kaplun, B. J. and Moreinis, J. M.,** Versuch der experimentellen Sensibilisierung gegen Salvarsan bei Menschen und Tieren, *Acta Derm.-Venereol.*, 11, 295, 1930.
330. **de Weck, A. L. and Frey, J. R.,** *Immunotolerance to Simple Chemicals*, S. Karger, Basel, 1966.
331. **Grolnick, M.,** Mechanism of spread of sensitization in eczematous contact dermatitis in humans, *J. Allergy*, 36, 333, 1965.
332. **Humphrey, J. H.,** The mechanism of Arthus reactions. II. The role of polymorphonuclear leucocytes and plateletels in reversed passive reactions in the guinea-pig, *Br. J. Exp. Pathol.*, 36, 283, 1955.
333. **Willoughby, D. A., Walters, M. N., and Spector, W. G.,** Lymph node permeability factor in dinitrochlorobenzene skin hypersensitivity reaction on guinea pigs, *J. Immunol.*, 8, 578, 1965.

334. **Pick, E. and Feldman, J. D.**, Transfer of cutaneous hypersensitivity to tuberculin in the guinea pig by γG_2 from immunized donors, *J. Immunol.*, 100, 858, 1968.
335. **Dvorak, H. F. and Mihm, J. C., Jr.**, Basophilic leukocytes in allergic contact dermatitis, *J. Exp. Med.*, 135, 235, 1972.
336. **Gell, P. G. H. and Hinde, I. T.**, The histology of the tuberculin reaction and its modification by cortisone, *Int. Arch. Allergy Appl. Immunol.*, 5, 23, 1954.
337. **Flax, M. H., Elliot, J. H., Daly, J. J., Willms-Kretschmer, K., McCarthy, J. S., and Leskowitz, S.**, Local plasmacytopoiesis in delayed hypersensitivity reactions, *J. Immunol.*, 102, 1214, 1969.
338. **Battisto, J. R.**, Adoptive cutaneous anaphylaxis: a technique for studying antibody globulin synthesis, *J. Immunol.*, 97, 939, 1966.
339. **Battisto, J. R.**, Spontaneous delayed iso-hypersensitivity in guinea pigs, *J. Immunol.*, 101, 743, 1968.
340. **Follett, D. A. and Battisto, J. R.**, Iso-Antibodies to a β-globulin that detects spontaneous delayed iso-hypersensitivity in guinea pigs, *J. Immunol.*, 101, 753, 1968.
341. **Katz, D. H., Paul, W. E., Goidl, E. A., and Benacerraf, B.**, Carrier function in anti-hapten antibody responses. III. Stimulation of antibody synthesis and facilitation of hapten-specific secondary antibody responses by graft versus host reactions, *J. Exp. Med.*, 133, 169, 1971.
342. **Metaxas, M. N. and Metaxas-Bühler, M.**, Passive transfer of local cutaneous hypersensitivity to tuberculin, *Proc. Soc. Exp. Biol. Med.*, 69, 163, 1948.
343. **Metaxas, M. N. and Metaxas-Bühler, M.**, Studies on the cellular transfer of tuberculin sensitivity in the guinea pig, *J. Immunol.*, 75, 333, 1955.
344. **Blazkovec, A. A., Sorkin, E., and Turk, J. L.**, A study of the passive cellular transfer of local cutaneous hypersensitivity. I. Passive transfer of delayed hypersensitivity in inbred and outbred guinea pigs, *Int. Arch. Allergy Appl. Immunol.*, 27, 289, 1965.
345. **Jansen, L. H. and Bleumink, E.**, Flare and rash reactions in contact allergy of the guinea-pig, *Br. J. Dermatol. Suppl.*, 83, 48, 1970.
346. **Bleumink, E. and Jansen, L. H.**, Studies on flare and rash phenomena in guinea-pigs, *Arch. Dermatol. Forsch.*, 242, 285, 1972.
347. **Cannell, H.**, The flare reaction and its rash component in guinea-pig oral mucosa and skin, *Br. J. Exp. Pathol.*, 53, 390, 1972.
348. **de Weck, A. L., Frey, J. R., and Geleick, H.**, Specific inhibition of contact dermatitis to dinitrochlorobenzene in guinea pigs by injection of haptens and protein conjugates, *Int. Arch. Allergy Appl. Immunol.*, 24, 63, 1964.
349. **Fukushiro, S., Nakagawa, S., Gotoh, M., Koshizawa, M., and Tanioku, K.**, The distribution of antigen in flare up reaction in contact sensitivity to DNCB, *Immunology*, 34, 549, 1978.
350. **Grolnick, M.**, Contact allergy of the skin, *Ann. N. Y. Acad. Sci.*, 50, 718, 1949.
351. **Dienes, L. and Simon, F. A.**, The flaring up of injection site in allergic guinea pigs, *J. Immunol.*, 28, 321, 1935.
352. **Skog, E.**, Spontaneous flare-up reactions induced by different amounts of 1,3-dinitro-4-chlorobenzene, *Acta Derm.-Venereol.*, 46, 386, 1966.
352a. **Turk, J. L.**, unpublished observation.
353. **Skog, E. and Wahlberg, J. E.**, Generalized dermatitis with epidermal eosinophilia induced in dinitrochlorobenzene-sensitized guinea pigs, *Dermatologica*, 143, 209, 1971.
354. **Skog, E. and Wahlberg, J. E.**, Immunological tolerance and dermal eosinophilia induced in DNCB-sensitized guinea pigs, *Acta Derm.-Venereol.*, 54, 437, 1974.
355. **Shelmire, B.**, Nature of the excitant of poison ivy dermatitis, *Arch. Dermatol. Syphilol.*, 42, 405, 1940.
356. **Kligman, A. M.**, Poison ivy (Rhus) dermatitis, *Arch. Dermatol.*, 77, 49, 1958.
357. **Lewis, G. P.**, Bradykinin, *Nature (London)*, 192, 596, 1961.
358. **Coombs, R. R. A. and Gell, R. G. H.**, The classification of allergic reactions underlying disease, in *Clinical Aspects of Immunology*, F. A. Davis, Philadelphia, 1963, 317.
359. **Cohen, S. and Ward, P. A.**, In vitro and in vivo activity of a lymphocyte and immune complex-dependent chemotactic factor for eosinophils, *J. Exp. Med.*, 133, 133, 1971.
360. **de Weck, A. L.**, Immune responses to environmental antigens that act on the skin: the role of lymphokines in contact dermatitis, *Fed. Proc. Fed. Am. Soc. Exp. Biol.*, 36, 1742, 1977.
360a. **de Weck, A. L.**, unpublished results.
361. **Engelbrigsten, J. K.**, Some investigations on hypersensitiveness to bichromate in cement workers, *Acta Derm.-Venereol.*, 32, 426, 1952.
362. **Engelbrecht, H.**, Maurerekzem und Chromgehalt des Zements, *Hautarzt*, 3, 542, 1952.

363. **Bonnevie, P.**, Quicklime cement eczema and its combination with allergy to chromate, *Acta Derm.-Venereol.*, 32 (Suppl. 29), 53, 1952.
364. **Ramos et Silva, J.**, L'eczéma du ciment, in *Le mécanisme physio-pathologique de l'eczéma*, Masson & Cie, Paris, 1954.
365. **Lejhancová, M. and Wolf, J.**, The role of the potassium dichromate test in eczemas caused by cement, *Cesk. Dermatol.*, 29, 334, 1954.
366. **Bering, F. and Zitzke, E.**, *Berufliche Hautkrankheiten*, L. Voss, Leipzig, 1935.
367. **Sassi, C.**, Pathologie der Gewerbekrankheiten in der Chromatfabrikation, *Med. Lav.*, 47, 319, 1956.
368. **Tataru, C., Marinescu, A., Capusan, I., Ripan, R., and Liteanu, C.**, Hautschädigungen bei der industriellen Gewinnung von Kaliumbichromat, *Berufs-Dermatosen*, 5, 218, 283, 1957.
369. **Urchs, O.**, *Berufliche Hautkrankheiten*, Cantor, Aulendorf, West Germany, 1953.
370. **Wendelberger, J. and Weitgasser, H.**, Ueber die Bedeutung von Alkali und Entfettungsmitteln für das Zustandekommen von Kontaktekzemen bei Galvaniseuren, *Berufs-Dermatosen*, 4, 16, 1956.
371. **Schroeder, K. H.**, Nickelekzem bei einem Lichtpauser, *Berufs-Dermatosen*, 3, 170, 1955.
372. **Köhler, H.**, Konstitutionsuntersuchungen bei Hautkranken in der Offsetabteilung eines Druckereibetriebes, *Berufs-Dermatosen*, 4, 93, 1956.
373. **Nater, J. P.**, Possible causes of chromate eczema, *Dermatologica*, 126, 160, 1963.
374. **Schultheiss, E.**, *Gummi und Ekzem*, Cantor, Aulendorf, West Germany, 1959.
375. **Burckhardt, W.**, Die beruflichen Hautkrankheiten, in *Handbuch der Haut- & Geschlechtskrankheiten*, Vol. II, Jadassohn, J., Ed., Springer-Verlag, Berlin, 1962, 398.
376. **Sidi, E. and Longueville, R.**, Dermites au ciment. Rôle des sels de chrome, *Arch. Mal. Prof. Med. Trav. Secur. Soc.*, 14, 41, 1953.
377. **Calnan, C. D. and Harman, R. R. M.**, Studies in contact dermatitis, XIII. Diesel coolant chromate dermatitis, *Trans. St. John's Hosp. Dermatol., Soc.*, 46, 13, 1961.
378. **Guy, W. B.**, Dermatologic problems in the railroad industry resulting from conversion to Diesel power, *Arch. Dermatol.*, 70, 289, 1954.
379. **Kaplan, I. and Zeligman, I.**, Occupational dermatitis in railroad workers, *Arch. Dermatol. Syphilol.*, 85, 95, 1962.
380. **Winston, J. R. and Walsh, E. N.**, Chromate dermatitis in railroad employees working with diesel locomotives, *JAMA*, 147, 1133, 1951.
381. **Schwartz, L. and Dunn, J. E.**, Dermatitis occurring among operators of air-conditioning equipment, *Ind. Med. Surg.*, 11, 375, 1942.
382. **Samitz, M. H.**, Some dermatologic aspects of the chromate problem, *Arch. Ind. Health*, 11, 361, 1955.
383. **Samitz, M. H.**, Studies of chromate dermatitis, *Arch. Ind. Heath*, 14, 269, 1956.
384. **Spier, H. W.**, Holzkonservierungsmittel und Chromatekzem, *Desinfekt. Gesundheitswes.*, 4, 58, 1956.
385. **Anderson, F. E.**, Cement and oil dermatitis, the part played by chrome sensitivity, *Br. J. Dermatol.*, 72, 108, 1960.
386. **Gerauer, A.**, Zur wachsenden Bedeutung der epicutanen Chromallergie, *Z. Haut-Geschlechtsk.* 17, 97, 1954.
387. **Samitz, M. H. and Gross, S.**, Extraction by sweat of chromium from chrom tanned leathers, *J. Occup. Med.*, 2, 12, 1960.
388. **Samitz, M. H., Katz, S., and Gross, S.**, Nature of the chromium extracted from leather by sweat, *J. Occup. Med.*, 2, 435. 1960.
389. **Morris, G. E.**, Chromate dermatitis from chrome glue and other aspects of the chrome problem, *Arch. Ind. Health*, 11, 368. 1955.
390. **Bory, R. and Lecocq, I.**, Eczéma par cuir chromé, *Arch. Mal. Prof. Med. Trav. Secur. Soc.*, 14, 144, 1953.
391. **Boström, G.**, Ueberempfindlichkeitsprüfung bei einem Fall von Chromallergie, *Acta Dermatol.-Venereol.*, 17, 631, 1936.
392. **Baer, H.**, Dermatitis infolge Gebrauchs einer mit "Ersatzleder" versehener Bartbinde, *Muench. Med. Wochenschr.*, 67, 874, 1920.
393. **Folesky, H.**, Ueber die Häufigkeit der subklinischen Chromatallergie in verschiedenen Berufen im Vergleich zu beruflich nicht Exponierten, *Symp. Dermatol. Morbis Cutaneis Professionalibus*, 1, 261, 1962.
394. **Levine, B. B. and Benacerraf, B.**, Studies on antigenicities. The relationship between in vivo and in vitro enzymatic degradability of hapten-polylysine conjugates and their antigenicities in guinea pigs, *J. Exp. Med.*, 120, 955, 1964.

Chapter 5

ADVERSE CHROMATE REACTIONS ON THE SKIN

Desmond Burrows

TABLE OF CONTENTS

I.	Chrome Ulcers	138
II.	Prevention of Chrome Ulcers	139
III.	Treatment	140
IV.	Chromate Sensitization Skin Reactions	140
V.	Chromate Patch Testing	141
	A. Dilution of Materials used in Patch Tests	141
	B. Trivalent or Hexavalent Chromium	142
	C. Type of Reaction	142
	D. Site of Testing	142
	E. Significance of Patch Tests	142
	F. Persistance of Positive Patch Tests	143
	G. Percentage Positive	143
	H. Length of Application of Patch Test	143
	I. Incidence	143
VI.	Patterns of Dermatitis	144
VII.	Associated Light Sensitivity	146
VIII.	Prognosis	146
IX.	Causes of Chromate Contact Dermatitis	146
X.	Cement	147
	A. Historical	147
	B. Current Trends	147
	C. Causation	147
	D. Chromate in Cement	148
XI.	Antirust/Corrosive	149
	A. Primer Paints	149
	B. Galvanizing	150
	C. Chromium Trioxide	150
	D. Antirust Agent in Coolant	150
	E. Defatting Solvents	151
	F. Brewery	151
XII.	Other Causes	151

	A.	Welding 151
	B.	Leather 152
	C.	Pigment 152
	D.	Printing 153
	E.	Glues 154
	F.	Ashes 154
	G.	Foundry Sand 154
	H.	Matches 154
	I.	Wood and Paper Industry 154
	J.	Machine Oils 154
	K.	Timber Preservatives 155
	L.	Boiler Linings 155
	M.	Television Workers 155
	N.	Magnetic Tapes 155
	O.	Tire Fitters 155
	P.	Chromium Plating 156
	Q.	Milk Testers 156
	R.	Food Laboratory 156
	S.	Bleaches and Detergents 156
		1. Bleaches 156
		2. Detergents 156
XIII.	Prevention of Chromium Dermatitis ...157
	A.	The Effect of Oral Chromium on Dermatitis ...157

References 158

I. CHROME ULCERS

Chrome ulcers were first described by William Cumin,[1] in Glasgow 1827, in two dyers and in a person involved in the manufacture of potassium bichromate. He used potassium bichromate locally as a treatment for tuberculosis and found this produced papules which became pustules and ulcerated. It has since been well-recognized as a hazard in tanners,[2] electroplaters,[3] and workers who manufacture bichromate.[4] It is the most common lesion resulting from occupational exposure to chromium.[5] In Britain, the majority of cases occur in electroplaters.[6] The lesions begin as single or multiple, painless, papular lesions on the hands, forearms, or feet, and are often ignored until the surface ulcerates with a tenacious crust which, if removed, leaves a punched out ulcer 2 to 5 mm in diameter with an undermined border. The ulcers can penetrate deeply, become painful, and reach the underlying bone (which it does not usually invade, though joints and cartilage can be penetrated). Most investigators believe that chrome ulcers do not occur on intact skin, hence the distribution on areas subject to trauma. Samitz and Epstein[7] found that trauma was essential to produce the lesion in experimental work in guinea pigs. Unless treated early, healing is slow with an atrophic scar. Neoplastic changes never occur. A large percentage have associated nasal ulceration; Edmondson[8] found 133 (76%) of workers with nasal ulceration had associated

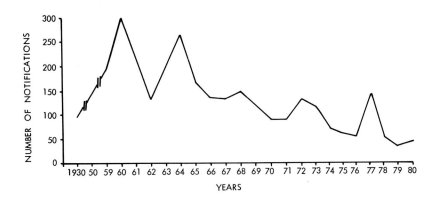

FIGURE 1. Notification of chrome ulcers in U.K.

skin ulceration. The incidence in an industry is directly related to the amount of exposure,[6] and in the U.K. there has been a statutory duty to report cases since 1919. Figure 1 shows that chrome ulceration of the skin is still a problem, even though there is probably considerable underreporting. In a retrospective study of 997 platers,[6] it was found that 21.8% had skin ulcers, but when the hygienic conditions were poor it could rise to 34%. Krishna et al.,[9] in India, found the incidence rose to 61.6% with poor conditions. Chromic acid and hexavalent chromates cause the lesion; probably chromic acid is toxic because of its oxidizing potency rather than its acidity.[10]

Edmondson,[8] in a survey of a large chemical plant that processes chromium ore to usable chromium chemicals, found that 198 of 285 men examined had either chrome ulcers or scars. Distribution was 46.3% on the arm, 19.5% on the hands, 10.1% on the ankles and feet, and other areas affected in descending order of frequency were legs, back, knees, thighs, abdomen, face, neck, and chest.

Patients with chrome ulceration do not exhibit allergy to chromates. Edmondson,[8] on patch testing 56 patients with chrome ulcers, found only two who had positive patch tests to 0.5% potassium bichromate but both of these had a history of dermatitis as well.

II. PREVENTION OF CHROME ULCERS

1. The introduction of the recently developed plating process in which chrome is deposited from trivalent chrome. Trivalent salts present a lesser toxic hazard than the traditional hexavalent chromic acid plating process.[11]
2. A high standard of cleanliness and protective clothing.
3. Low atmospheric levels. Good equipment design with strong exhaust ventilation at tank level and antimisting agents.
4. Skin cuts and abrasions, however slight, must be cleaned immediately with water and then a 10% $CaNa_2$ EDTA ointment (calcium disodium salt of ethylemidiamine tetra acetic acid) should be applied. EDTA does not chelate hexavalent chromium at room temperature but it reduces the hexavalent form to the trivalent form rapidly and the excess EDTA chelates the trivalent chromium.[5] In experimental work in guinea pigs,[12] it was found that 5 and 10% ascorbic acid had a protective effect when added to 20% dichromate solution but a 1% solution provided no protection. An "antichrome" powder (40% sodium metabisulfite, 27% ammonium chloride, 20% tartaric acid, 20% sucrose, prepared in a 10% aqueous solution), which reduces hexavalent to trivalent chromium which then complexes with the tartaric acid, did

not have a protective effect unless applied within minutes of contact with chromium.[7]

The present U.K. chrome plating regulations are as follows:

1. Tests must be carried out every 14 days to ensure that environmental levels are not above the following: soluble chromic, chromous salts—Cr-0.5 mg/m^3 of air; chromates, dichromates, chromic acids—0.1 mg/m^3 of air. Chromium metal and insoluble salts—10.0 mg/m^3 of air excluding hexavalent chromium compounds.
2. Regular inspection of hands and forearms.
3. Various preventive measures—gloves, washing facilities, and application of an ointment containing soft white paraffin (three parts) and lanolin (one part) to the uncovered parts of the skin.

III. TREATMENT

If removed from contact with chromium, the ulcers will heal in several weeks,[7] Dewirtz[13] reported the average healing rate in 297 cases was 20 days as an out-patient and 10 days as an in-patient. Samitz and Epstein,[7] in experimental lesions in guinea pigs, did not find any increase in speed of healing with an effective antichrome agent. Maloof[14] treated 54 chrome skin ulcers in the tanning industry with an ointment containing 10% EDTA applied under a dressing which was removed in 24 hr and the ulcer crater cleared out. In almost every case it was found that the base of the ulcer was loose and easily removed and healing rapidly occurred. This, however, was not a controlled trial. Pirozzi et al.,[15] found that another reducing agent (ascorbic acid 10% aqueous solution) in experimental work in guinea pigs reduced the healing time of potassium dichromate treated ulcers to a third. An improved healing time was found even with a delay up to 30 min in applying the ascorbic acid solution. Sodium dithionite aqueous solutions (3 g/l) is commonly used in the electroplating industry to prevent ulcers. The solution is made up freshly each day and splashed on contaminated skin. It has the advantage over sodium metabisulfite in that it does not require an acid media to convert hexavalent to trivalent chromate.[16]

IV. CHROMATE SENSITIZATION SKIN REACTIONS

Chromium Valences[17]

Cr^0 Metal in alloys (e.g., stainless steel) and in plating, not soluble, not sensitizing
Cr^{3+} Salts of inorganic acids (e.g., chlorides), soluble, sensitizing, may precipitate in alkaline environment
   Salts of organic acids (e.g., oxalate), soluble, sensitizing, basic sulfate (leather tanning agent), soluble, sensitizing, oxide (Cr_2O_3), not soluble, not sensitizing
   Hydroxides, when aged not soluble, and then not sensitizing
   Complexes used in textile dyes, commonly not sensitizing
Cr^{4+} Dioxide (in magnetic tapes) forms Cr^{6+} and Cr^{3+} in presence of water, sensitizing.
Cr^{6+} Chromates, dichromates of K-,Na-,Ca-,NH_4-, dichromate of Zn, soluble, sensitizing
   Chromate of Zn, chromate and dichromate of Pb, weakly soluble, weakly sensitizing

The sensitizing capacity of chromate in a compound depends on the amount present, valency, solubility, pH, and presence of organic matter. Hexavalent soluble salts in an alkaline medium have the greatest sensitizing potential. Organic material has the potential to reduce hexa- to trivalent chromate.

Table 1
THRESHOLD OF SENSITIVITY TO DICHROMATE

	\multicolumn{6}{c}{Concentration (%)}	
	0.5	0.2	0.1	0.005	0.01	0.001	Ref.
	—	—	2	8	5	—	18
	10	7	7	—	5	2	19
	49	—	35	—	13	—	20
	1	4	25	—	4	1	21
	—	—	23	25	2	—	22
Total	60	11	92	33	29	3	

V. CHROMATE PATCH TESTING

It is essential to understand that allergy to chromium metal does not occur as in nickel sensitivity, hence potassium dichromate is the standard material for patch testing sensitivity to chromate compound ion.

A. Dilution of Materials used in Patch Tests

The results of patch testing with various dilutions is shown in Table 1. Dilution is critical in chromate patch testing. Allenby and Goodwin[23] re-patch-tested 14 patients, with a positive patch test to 0.5% potassium dichromate with varying dilutions and obtained the following result:

0.5%	Potassium dichromate—14 patients
0.25%	Potassium dichromate—10 patients
0.025%	Potassium dichromate— 3 patients
0.0025%	Potassium dichromate— 2 patients
0.00025%	Potassium dichromate— 1 patient

It is possible that 0.5% dilution is near irritancy level which may explain the lower incidence on re-patch-testing with a single chromate patch test, when the skin condition is quiescent, or no other skin condition present, e.g., light sensitivity.

Epstein,[24] suggested that some cases of chromate sensitivity may be missed by epicutaneous tests and that these can be demonstrated by intradermal testing and he quoted three illustrative cases. Metanic,[25] found only 24.5% of cement dermatitis cases showed positive patch tests to potassium dichromate, 0.05%, whereas 90% reacted to 0.0001% dilution intradermally. These figures are a little unusual as most investigators find 80 to 95% of cases of cement dermatitis patch test positive. Others[25-28] have suggested that the sensitivity of patch testing should be increased by adjusting the pH to 10 to 12 to correspond with the alkalinity of cement. Skog and Wahlberg,[28] found the mean sensitivity threshold in an alkaline buffer solution pH12 was 0.08%, in petrolatum 0.15%, and in water 0.25%, and they thought that this increased sensitivity may be due to increased skin penetration. Burrows and Calnan,[19] however, were unable to show any real difference between patch tests using chromate in an alkaline buffer and in a water solution. Rudzki et al.[29] attempted to increase the sensitivity of patch testing by the addition of emulsifiers (10% glyceryl monostereate, 10% Tween® 60, and 3% Span® 20) to facilitate release of chromate. Glyceryl monostereate was the only one which enhanced the release of chromate from bases with low concentration and improved the patch test sensitivity.

B. Trivalent or Hexavalent Chromium

Patch testing with trivalent chromium does not produce such a high percentage of positives as hexavalent chromium. Anderson[18] found that none of 19 patients who gave a positive patch test to potassium dichromate were positive to chromium chloride. Engelhardt and Mayer[30] found that none of their chromate dermatitis patients in the lithography industry, who were patch test positive to 0.5% potassium dichromate, reacted to trivalent chromium on patch testing. Fregert and Rorsman,[31] however, were able to elicit positive reactions to trivalent chromium on patch testing in 11 out of 17 patients when using 0.5 M chromium trichloride, but only 4 out of 22 when using 0.07 M. The patch test activity of trivalent compounds compared with hexavalent chromate is of the order 1/10 for the oxalate, 1/100 for the chloride, and 1/1000 for the acetate[32] (all trivalent compounds). Forstrom et al.[33] found that only 23 out of 46 patients who were positive to dichromate gave a positive reaction when tested with 1 to 5% solution of trivalent chromium. There was a higher percentage of reactions to trivalent chromate in those with dermatitis of the feet. Tests with 20% basic chromium sulfate are often positive.[34] Allenby and Goodwin[23] found that only 1 of 10 patients who were patch test positive to 0.25% potassium dichromate was positive to 0.25% chromic chloride. Rudzki et al.[35] were able to demonstrate that the difference in positive reactions between hexavalent and trivalent chromium was not due to any difference in release from the vehicle.

Hexavalent chromium (chromate and bichromate) and trivalent chromium are allergenic compounds. Hexavalent chromium penetrates more easily into the skin and is of greater importance as a cause of dermatitis. Probably hexavalent chromium is reduced to the trivalent form in the skin when a hapten is conjugated with protein to form an antigen. Hexavalent chromium does not combine with protein via a covalent bond but trivalent chromium does. Hexavalent chromium compounds are more soluble than trivalent chromium compounds. Trivalent compounds are used in tattoos and hence tattooing skin reactions infrequently give allergic reactions.

C. Type of Reaction

A positive patch test will consist of redness with some induration; severe reactions will produce a definite eczematous response. Many metals, e.g., nickel, cobalt, and chromate can produce pustular reactions which do not indicate allergy.

D. Site of Testing

The site of patch testing is important. Cronin[36] thought that the back gave more positive reactions than the arm; probably the skin of the back is not more reactive than the arm but penetration may be greater. Nielsen and Kalaschka[37] showed that the skin was a 5 to 12 times more sensitive indicator than the mucosa.

E. Significance of Patch Tests

Positive patch tests to potassium dichromate have been found in 2 to 5% of healthy cement workers.[19,38,39] In Burrows and Calnan's report[19] of patch testing 70 healthy cement workers, two gave a positive patch test. It is interesting that one of these developed dermatitis 6 months later. Walsh[40] found three positive patch tests to 1% potassium dichromate among 60 patients with no skin disease in a tuberculous sanatorium, though at this percentage these may have been irritant reactions. In surveys of patients considered to have eczema with no obvious contact with chromium, positive patch tests were obtained with concentrations of chromate varying from 0 to 12%.[18,30,39,41–43] Chronin[44] found that only 40% of 48 men allergic to chromate were diagnosed as chromate sensitive prior to patch testing.

Table 2

Areas	Patients tested	+ Chromate %	Ref.
Australia	1000	13.5	46
Belgium	400	14.0	47
Brazil	536	15.7	48
California	223	9.4	49
Denmark	3225	5.7	50
Europe	4825	6.6	51
England	932	6.4	51
Holland	378	5.3	51
Italy	315	20.6	51
Ireland	529	5.3	52
Kuwait	389	57.0	53
New York	769	11.6	54
North America	1200	8.0	55
Poland	1205	16.2	56
		13.4	57
Scandinavia	5558	7.4	58
Scotland	1312	12.0	59
Spain	2806	17.5	60

F. Persistence of Positive Patch Tests

Thormann et al.[45] re-patch-tested 48 patients with a positive patch test reaction to 0.5% potassium dichromate. After 4 to 7 years, 38 patients (79%) still had a positive patch test.

G. Percentage Positive

The percentage of patients from various countries suspected of contact dermatitis from any cause found to show a positive patch test to 0.5% potassium dichromate is shown in Table 2. The variation in different areas probably depends on the percentage of the population engaged in the building industry, for instance, Fregert et al.[51] found a high incidence (20.6%) in Southern Italy where many of the people tested were men in the construction industry. Similarly, Moriearty et al.[48] in Brazil and Kanan[53] in Kuwait considered that their higher incidence was due to the high percentage of the population employed at the time of testing in the construction industry in hot humid conditions. The percentage of positives has remained fairly constant in a given area over a period of years.[50,61-63]

H. Length of Application of Patch Test

The standard time for leaving the potassium dichromate patch test on the skin is 48 hr with reading of the patch test on removal and 48 hr later. Some dermatologists claim that the patch test can be removed after 24 hr without impairment of the results.[63,64] Rudzki et al.[64] found that out of 30 patients positive to 0.5% potassium dichromate when the patch test was applied for 48 hr, 28 were also positive when the patch test was removed after 24 hr. He found that 90.1% of the potassium dichromate was released from petrolatum within the first 24 hr.

I. Incidence

In women, the occupations were as follows: housework 23%, domestic cleaning, engineering, leather work, cement, printing, and those handling primer paints, artists paints, and cement. No source was found in 16% of patients. The reverse of the normal

Table 3
SEX INCIDENCE OF CHROMATE SENSITIVITY

Area	+ Chromate		Sex ratio (M:F)	Ref.
	M	F		
Scandinavia	11.7%	3.0%	4:1	58
Scandinavia	10.0%	2.0%	5:1	66
Europe	10.7%	3.6%	3:1	51
N. America	10.0%	6.0%	2:1	55
Kuwait	218 patients	3 patients	73:1	67
England	519 patients	163 patients	3:1	68
Ireland (1981)	9 patients	1 patient	9:1	Table 6
Israel	9 patients	43 patients	1:5	69
Scotland	7.0%	4.0%	1.7:1	59
Poland	131 pts	119 pts	1.1:1	57
Brazil	25.5%	5.7%	4.4:1	48
Belgium	17.3	9.7	1.8:1	47

male:female ratio in Israel is attributed to contact with chromates in detergents and bleaches (see Section XII.S).

The higher incidence in men reflects their occupational contact (Table 3). Cronin[70] found 46% of her cases of chromate dermatitis were due to cement. Other common male occupations involving contact with sensitization to chromate were engineering, printing, plating, and leather tanning.

There were 28 positive patch tests to 0.5% potassium dichromate out of 731 patients tested with the Trolle Lassen battery at the Skin Department, Royal Victoria Hospital, Belfast, from April 1980 to March 1981. Of the 28 positive, 25 returned for re-patch-testing with a single 0.5% potassium dichromate patch test. The results are shown in Tables 4, 5, and 6.

These figures indicate that it is essential to re-patch-test, with a single patch, patients who have a positive dichromate test, particularly if the subject is female and has other positives. These false positives are probably a manifestation of the "angry back syndrome,"[74] where a strong positive to one substance will produce other positive tests nearby. It is probably uncommon to get true concurrent associated nickel and cobalt allergy. This phenomenon, together with the number of false positives in patients with eczema, suggest that 0.5% potassium dichromate may not be far from irritancy levels.

VI. PATTERNS OF DERMATITIS

The majority of cases of contact dermatitis to chromate follow the same pattern for any other materials involving the finger webs, backs and sides of fingers, backs of hands, and fronts of wrists. An appreciable percentage, however, have atypical patterns, often resembling constitutional eczema. Shanon[70] described 14 patients with a rash indistinguishable from atopic dermatitis who gave a positive patch test to 0.5% potassium dichromate and who improved on removal from contact with chromate. Engel and Calnan[71] reported 45 cases of chromate dermatitis due to primer paint. All of these had the typical distribution of chrome dermatitis in the hands but the eruption varied in its clinical appearance and distribution over weeks and months, and occasionally was virtually indistinguishable from atopic, nummular, or seborrhoeic eczema. Hall[72] and Bernhardt[73] similarly noted the nummular pattern in patients with chromate dermatitis. In Burrows and Calnan's[19] study of 134 patients with cement dermatitis,

Table 4
MALE PATIENTS WITH POSITIVE PATCH TEST TO POTASSIUM DICHROMATE 0.5%—ROYAL VICTORIA HOSPITAL, 1980, RE-PATCH-TESTED WITH SINGLE POTASSIUM DICHROMATE 0.5% PATCH, 6 TO 12 MONTHS LATER

Occupation	Diagnosis	Other positive patch tests	2nd Testing
Sales representative for masonry guns, works beside cement, was machinist	Eczema	—	++
Plasterer	Contact dermatitis	—	++
Flagger	Contact dermatitis	Cobalt and nickel	—
Bricklayer	Contact dermatitis	—	++
Fitter—Aircraft—? Primer Paint	Contact dermatitis	—	++
Builders laborer	Contact dermatitis	—	++
Photographer, do it yourself, cement, shoes	Contact dermatitis	—	++
Plasterer	Actinic reticuloid and contact dermatitis	Epoxy resin	++
Machinist	Contact dermatitis and seborrhoeic eczema	—	++
Bricklayer	Contact dermatitis	—	++
Bricklayer	Contact dermatitis	Rubber, Mercapto-mix Parabens	Did not attend
Microbiologist	Contact dermatitis and seborrhoeic eczema	—	Did not attend

Table 5
FEMALE PATIENTS WITH POSITIVE PATCH TEST TO POTASSIUM DICHROMATE 0.5%—ROYAL VICTORIA HOSPITAL, 1980, RE-PATCH-TESTED WITH SINGLE PATCH POTASSIUM DICHROMATE 0.5%, 6 TO 12 MONTHS LATER

Occupation	Diagnosis	Other positive patch tests	2nd Testing
Housewife	Varicose eczema	Neomycin, nickel, and cobalt	—
Housewife	Contact dermatitis	Nickel and cobalt	—
Housewife	Eczema	—	—
Housewife	Contact dermatitis	Nickel and cobalt	—
Housewife	Contact dermatitis	Mercapto-mix and Rubber	—
Housewife	Contact dermatitis	Balsam of Peru, fragrance mix, colophony, nickel, and cobalt	—
Housewife	Leg ulcer	Parabans and lanolin	—
Prison warden	Atopic eczema	Nickel, Boots E45 cream	—
Clerk	Atopic eczema	—	—
Teacher	Atopic eczema	Cosmetics	—
Draughts person	Contact dermatitis	Thiuram-mix, Carba-mix, Nickel	+
Drawing office	Eczema	Nickel	Did not attend
Sales assistant	Contact dermatitis	Nickel and cobalt	—
Burser	Contact dermatitis	Cobalt	—
Clerical assistant	Contact dermatitis	Perfume	—

Table 6
PATIENTS WITH POSITIVE PATCH TEST TO DICHROMATE 1980, ROYAL VICTORIA HOSPITAL, BELFAST, ON TROLLE LASSEN BATTERY, RE-PATCH-TESTED WITH SINGLE DICHROMATE PATCH TEST 0.5%

	Total patients	Female	Male
Total chromate patch test positive	28	16	12
Retested	25	15	10
Positive on retesting	10	1	9
Re-patch-tests positive only to potassium dichromate (first visit)	8	1	7

13% of the patients presented with discoid eczema and 10% with seborrhoeic eczema involving the face, scalp, ears, neck, axilla, waist, and groin. The dermatitis simulated stasis eczema in 7%. Eczema of the palms occurred in 9%.

VII. ASSOCIATED LIGHT SENSITIVITY

Frain-Bell[75] believes that allergic sensitivity to potassium dichromate is commonly found in photosensitivity dermatitis and actinic reticuloid syndrome. Wahlberg and Wennersten[76] showed that a lower concentration of dichromate will produce positive patch tests when irradiated with UV Light B but not with UV Light A. Feuerman[77] considered that he saw a high incidence of chromate dermatitis in women in Israel because the sunlight facilitated sensitization. It is doubtful, however, if there is a connection between photosensitivity and chromate allergy. The concentration of 0.5% used may be near irritancy level and may produce positive patch tests in persons suffering from extensive skin disease.

VIII. PROGNOSIS

The medical and social fate of the dichromate allergic patient has been well-reviewed by Breit and Turk.[78] Chromate dermatitis has a bad prognosis by the criteria of continuation of dermatitis, social and financial fate and work prospects.[79] In a 10 year follow-up of occupational dermatitis, Fregert[80] found that only 7% (mean for all causes 16%) of women and 10% (mean for all causes 24%) of men were healed. Purschel and Furst[81] found that a large percentage of their cases were still in contact with chromate in spite of a change of occupation. Burrows[79] found that only 8% of his cases of cement dermatitis (in a medico-legal survey) were cleared 10 to 14 years later. Younger patients have a better prognosis.[79]

IX. CAUSES OF CHROMATE CONTACT DERMATITIS

Parkhurst[82] was probably the first to report chromium dermatitis when he described a woman working with potassium dichromate in blue print production who developed dermatitis and had a positive patch test to 0.5% solution of potassium dichromate. Smith[83] reported a man with dermatitis, working with ammonium dichromate who reacted to a 1% ammonium dichromate patch test. However, credit is generally given to Engelhardt and Mayer[30] for first establishing the connection between chromate sensitivity and dermatitis when investigating dermatoses in lithographers. In the 114 workers studied, 30 had eczema and of these, 84% reacted to 0.5% potassium dichromate applied as a patch test.

X. CEMENT

A. Historical

Cement has long been known as a cause of contact dermatitis. Bernadino Ramazzini referred to it in his book, *De Morbis Artificia Diatriba,* first published in 1700, when he noted "lime makes the hands of bricklayers wrinkle and sometimes ulcerates them."

Charles Turner Thackrah, whose book was published in 1831, referred to agents injurious to skin such as lime, which produced cutaneous disease in bricklayers. Modern cement manufacture is usually credited to Aspdin in England about 1824 and dermatitis has been described in almost every country since.[84] O'Donovan in 1925,[85] in England, saw many men affected by cement while working in the tunneling for the London Underground Railway System and he noted the "constitutional eczema" patterns. Cement was made in vertical furnaces till about 1900 and these may not have produced hexavalent chromium. The connection between chromate sensitivity and cement workers was first described in 1939 by Bonnevie[86] and Stauffer,[87] but it was thought at the time to be due to chromium compounds in leather in gloves. Pirila and Kilpio,[88] reported that 10 sufferers from cement dermatitis were sensitive to chromate. Jaeger and Pelloni[42] demonstrated that 32 cement eczema patients with cement dermatitis all had a positive patch test to potassium dichromate but of the 168 other patients who were tested and who had eczema from other causes, only 5% reacted. So in 1950, they first proved the connection between cement dermatitis, chromium allergy, and cement by demonstrating quantities of chromium in cement as well as chromate sensitivity of their patients.

B. Current Trends

Cement is the most common cause of primary sensitization to chromate.[89,90] Recently, however, there has been a decrease in a number of cases of cement dermatitis. Burrows,[52] in a review of 529 patients having routine Trolle Lassen battery patch tests, found 28 (5.3%) with positive patch test to chromate but in only 4 was there any possible contact with cement. This reduction is probably due to changing patterns of handling cement, i.e., using readyuse cement, and the provision of greater mechanization on building sites, as well as greatly improved facilities for personal hygiene of the employees.

C. Causation

Cement can cause dermatitis both by direct irritation or sensitization to chromate or both. The fact that cement is alkaline, abrasive, and hydroscopic is probably the reason why more men are allergic to chromate in cement than through contact with other sources which have an equal amount of chromate present in them. Calnan in 1960[84] made the observation that dried cement is relatively innocuous and very few cases of cement dermatitis were ever seen in a cement making factory. General experience confirms this observation. Wet cement is much more alkaline than dry cement because water liberates calcium hydroxide and this so called scaling of lime causes a marked rise in pH and the mixture with sand is abrasive. Rabito and Peserico[91] suggested that wet cement had an oxidizing capacity not present in dry cement and that this was partly responsible for its irritancy.

Sensitization to chromate is, however, of more importance than irritancy in causing cement dermatitis (Table 7).

Table 7
CORRELATION OF CEMENT ECZEMA WITH CHROMATE ALLERGY (HOVDING-amended)[97]

Number of cases	Positive reactors	%	Ref.
42	39	93	38
66	63	95	18
60	31	52	41
171	134	78	19
83	78	94	92
8	7	88	93
33	33	100	94
100	95	95	95
14	10	71	96
20	17	85	97
60	55	92	39
32	30	94	42
28	20	76	98
35	1	3	99
246	187	76	100
17	13	76	26
	Average	80.1	

Table 8
ANALYSIS OF THE INGREDIENTS IN CEMENT

Material	Chromate expressed as ($K_2Cr_2O_7$) µg %	Total chromium µg %
Gypsum	0	2,920
Chalk	0	740
Clay	0	21,840

D. Chromate in Cement

Portland cement is made from chalk or limestone and clay or shale with the addition of gypsum. Clay is mixed with water with crushed chalk or limestone added, and this is then progressively milled in a ball or roller mill using chrome-steel. It is then passed through a kiln at a temperature of 1200 to 1400° C, heated by coal or oil, some of which may be incorporated in the cement. This kiln is lined with chromium containing refractory bricks. After the kiln stage, gypsum is added and further ground. Johnston and Calnan[101] analyzed the ingredients in cement (Table 8).

The possible sources of chromate in the finished product are (1) the ingredients—clay, gypsum, and chalk; (2) refractory bricks; (3) chrome steel grinders; and (4) ash. Trivalent chromate from these sources can be converted to hexavalent chromate in the kiln.

Johnston and Calnan[101] observe that 130 parts of cement are produced from 120 parts of chalk, 40 parts of clay, 5 parts of ash, and 5 parts of gypsum; about 40 parts of total are lost at process and manufacture. On this basis, 70% of chromate in cement would come from clay, 13% from coal ash, 7% from chalk, and 11% from other sources.

Table 9
INCIDENCE OF CEMENT DERMATITIS AS A PERCENTAGE OF INDUSTRIAL DERMATOSES

Country	Percentage of industrial dermatoses(%)	Ref.
Germany	10	26
Finland	12	88
France	20–25	103
	33	104
Switzerland	30.8 (1951)	105
	40.3 (1959)	105
	34.8 (1960)	105
Italy	32	106
Northern Ireland	23	79

Table 10
INCIDENCE OF CEMENT DERMATITIS IN CEMENT WORKERS

Country	%	Ref
Italy	1.3	106
Sweden	4.9 (Bricklayers)	107
	5.9 (Concrete workers	107
Norway	5.5–7	97
Northern Ireland	3.5	79
Australia	5.2	108

In a study in a cement factory over a period of time the variation of water soluble chromium was small. In the same study,[101] they showed that due to lack of oxygen and rotation in vertical furnaces, all chromate is in the trivalent form. Unfortunately, modern processing of Portland cement does not permit using vertical furnaces.

Hovding[97] reviewed 12 papers in which the chromate was analyzed. He found that it varied from 30 mg/kg hexavalent chromium and total chromium from 10,000 to 113,000 mg/kg. The quantity of soluble chromium measured can be affected by the method of determination and this could explain some of the variations quoted.

In Britain, the chromium extracted by water as soluble chromate from ordinary Portland cement is normally around 10 mg/kg, but may occasionally be as high as 30 mg/kg. The total chromium content tends to be around 60 mg/kg, but can range up to 120 mg/kg, though there does not appear to be any correlation between total and soluble chromium contents.

Cement dermatitis may take many years to develop. Hovding[97] found that of the 24 men who were patch test positive to dichromate, 14 had worked for 11 years or more before developing dermatitis. The incidence of cement dermatitis in cement workers and as a percentage of industrial dermatoses is shown in Tables 9 and 10.

XI. ANTIRUST/CORROSIVE

A. Primer Paints

There are three principal compounds used in primer paints: (1) zinc potassium chromate—this commonly used form contains zinc chromate and potassium dichromate,

(2) basic zinc chromate—zinctetraoxy-chromate, and (3) mixtures of these two types.

Zinc chromate is insoluble and basic zinc chromate slightly soluble. It is essential that primers should contain some soluble chromate (potassium dichromate) in order to inhibit corrosion. Adams et al.[109] concluded that zinc potassium chromate in primer paints contains water soluble chromate in amounts corresponding to patch test solutions and a much higher concentration than those found in cement. This would certainly explain the sensitizing capacity of these paints. Many authors have described cases of dermatitis to primers.[18,72,73,109,110] Chromate is now being removed from paints because of the carcinogenic risk in manufacture and few paints contain chromate, though there is no evidence of carcinogenesis in paint users.

B. Galvanizing

Good protection of iron against rusting can be obtained by galvinizing molten zinc. However, zinc can be oxidized leading to the formation of a white stain. Therefore, in recent years, metal and particularly galvanized zinc sheets have been coated with special hexavalent chromate compounds. Chromate is usually applied by spraying. After such treatment, the metal has a film consisting not only of practically insoluble zinc chromate and readily soluble chromate, but also trivalent chromium of less than 1 mu thickness. Fregert et al.[111] described this process in detail and gave details of five cases of contact dermatitis due to chromate in men handling galvanized sheets. They were able to detect 0.03 to 0.44 $\mu g/cm^2$ water soluble chromate from sheets, of which, 22 to 86% was hexavalent. Fregert et al.[111] studied seven workers without eczema who worked with galvanized sheets for 4 hr. Five of them washed their hands for 3 min in 1 ℓ of distilled water containing a few drops of chromium free liquid soap. Two of them washed their hands in water without soap. The total amount of chromium in content of hexavalent chromium was determined. In the water used in washing, of the five men who washed their hands with soap there was 1.04 to 5.00 mg per person, of this 42 to 51% was hexavalent. The amount of chromium recovered after washing in water without soap was 0.05 to 0.14 mg per person. This indicates the considerable chromate contamination of the hands and the value of washing with soap. In comparison, in three bricklayers after handling cement for 4 hr, no chromate was found in the hand washings of two of them and 0.007 mg in one case. Newhouse[112] described 47 cases of chromate dermatitis in car assemblers, where the dermatitis was traced to nuts, bolts, and screws which had been zinc plated and chromate treated (passivation). This process helps the parts to resist oxidation and other atmospheric erosion. Passivation is effected by the absorption on the surface of zinc of the chromate radicle which then becomes insoluble in cold water and allows subsequent rinsing of the part. Inadequate rinsing could result in additional chromate being retained on the surface increasing any hazards associated with the handling of chromated parts. Sodium dichromate (150 g/ℓ) is commonly used in the dip. This process is in addition frequently used in the aircraft industry.

C. Chromium Trioxide

About 40 g/ℓ is used in the aircraft industry to form a coating of aluminium oxide on the metal which should be chromium free, but there could be fumes of chromium in the atmosphere. Chromates are used for pretreating aluminium to ensure good adhesion of thick paint. No cases of dermatitis have been described, however. Recently, chromate has been incorporated in epoxy resins to prevent rusting in cracks in the resin.

D. Antirust Agent in Coolant

Chromates have been used for anticorrosion in many types of recirculating water since the early 1930s.[40] Winston and Walsh[113] reported six cases of dermatitis in men working in a diesel locomotive shop who were exposed to sodium dichromate which was used as an anticorrosion agent in coolant fluid. These patients were patch test positive to chromate. The formula for the powder was 66% sodium dichromate, 24% soda ash, 5% sodium phosphate, and 5% sodium silicate; 1.5 lb of this powder was dissolved in 2 gal of water producing 6% sodium dichromate solution. This concentrated solution was put into a radiator and filled with water giving a final concentration of 0.08% dichromate solution. Calnan and Harman[114] have fully reviewed the mechanism of corrosion and the place of chromate in its prevention and described six cases of chromate dermatitis due to contact with chromate in coolant water. Four of these were railway fitters; one was an electrician responsible for adding chromate powder to water circulating in a central heating system, and one a printer who worked on a printing machine which had a water cooling system. Plunkett,[115] Haeberlin,[116] and Guy[117] also described cases of chromate dermatitis due to chromate in coolant water. However, Marsland[118] reported there were no cases of chromate dermatitis in Western Australia when many locomotives were converted to diesel fuel and where chromate has been used in their cooling systems since 1949. He believed that this was due to careful use of the material in the liquid form only and a lower concentration than that used by many U.S. railway systems (65 grains/gal). Calnan[119] described a case of chromate dermatitis in a man whose job was to clean record presses. He was in contact with a coolant which contained between 269 and 357 ppm of chromate as an antirust agent. Chromates should be avoided as antirust agents in coolant fluids. Effective alternatives are available, such as nitrites and amines, but chromates have an additional use in coolants, which makes them more acceptable, in that calcium chromate is soluble and thus prevents calcium deposits in hard water areas.

E. Defatting Solvents

A case is described of chromate dermatitis due to 0.2% sodium chromate in a defatting solution used to clean engines.[120] The sodium chromate was present as an anticorrosion agent and remained on the surface of the engine so that it was possible that the skin of others workers who later assembled the engines could be contaminated.

F. Brewery

A case of chromate dermatitis in a worker who handled yeast residue has been described.[121] It was found that brine was added to yeast residues. The brine was composed of 560 parts of calcium chloride, 2 parts of sodium hydroxide, and 4 parts of sodium dichromate in 4000 parts water. The brine is used as a cooling agent to lower the freezing point of water and the chromate is added to prevent corrosion.

XII. OTHER CAUSES

A. Welding

Fregert and Ovrum[122] described a case of a welder who presented with repeated spells of facial dermatitis following exposure to welding fumes. The patient was found to be patch test positive to chromate. Traces of chromium have been found to be present as inclusions or as intentional constituents of the welding electrodes as well as rods used in oxygas welding. Some welding rods may contain as much as 18% chromium[123] though it should be remembered that only a minority of welding rods contain chromate. Part of the chromium may be oxidized to hexavalent chromium and carried off in the welding gas fumes.[124] Shelley[123] described a man who developed chromate dermatitis of his hands and who had repeated recurrences which were eventually traced to walking past an acetylene welding operation; when he stopped this, the condition cleared up.

Though industry has been well aware of the contents of welding fumes, few instructions are given to chrome sensitive patients.[124]

B. Leather

The leather tanning industry is a large user of salts of chromium. Chromate is used in leather for four purposes:[125]

1. Tanning—the objective of tanning being to stabilize collagen by the formation of chromium-collagen compound. Chromium in the form of hydroxyoxide chromium complex is thought to be associated with carboxylic groups on the collagen. A non-covalent bond is thought to create inter/intra chain cross-links. The most important chromium chemical used in leather manufacture is basic sulfate and trivalent chromium. Hexavalent chromium has not been used in tanning leather in the U.K. for 20 to 25 years. The commercial tanning compounds are usually added as concentrated salts containing either 11 or 15% W/V Cr_2O_3 or as a spray dried powder which contains 25% W/V Cr_2O_3.
2. Water repellant—trivalent chromium stearate chloride.
3. Stain repellants—trivalent chromium and fluorinated carboxylic acids form the basis of certain stain repellants.
4. Some dyes are based on coordinated complexes involving trivalent chrome and a mordant dye.

The chromium in chrome leather is insoluble in water but the chromium is extracted when lactic acid is present as in perspiration.[126] Samitz and Gross[127] confirmed this report and furthermore showed that extracts of leather by human sweat contained both hexavalent and trivalent chromium. The question as to whether leather can cause or exacerbate chromate dermatitis is controversial. Most authors would accept that contact dermatitis to shoes is unusual with people with proven cement dermatitis[19] though Rudzki and Kozlowska[57] found that 82.8% of their patients with contact dermatitis to chromate experienced exacerbation of their dermatitis by contact with leather. Morris[128] reported four patients with shoe leather dermatitis having positive patch tests to 0.2% basic chromic sulfate (trivalent). The incidence of chromate sensitivity in shoe dermatitis has been variously reported (Table 11).

Chromate sensitivity has been found to be an infrequent cause of leather dermatitis in Ireland (Tables 4, 5, and 6). In the year 1980 to 1981 there was one possible case of shoe dermatitis among ten cases of chromate dermatitis seen in Belfast. An appreciable proportion of leather, particularly from Asia, is vegetable tanned. Fregert and Gruvberger[138] showed that while this leather contained chromium it was in the same amounts as in animal tissue and was nonallergic. A worrying feature was that many of the gloves they tested, though labeled vegetable tanned, were clearly chrome tanned. Sometimes, vegetable tanned leather is retanned with chromium. The present position is that hexavalent and trivalent chromium can be leeched from leather and cause or exacerbate dermatitis. There is a high incidence of chromate dermatitis in tanneries despite the exclusive use of trivalent chromium. This is probably due to the effect of other irritants and maceration facilitating sensitization.

C. Pigment

Chromates are seldom used in the dye industry now with two exceptions in wool dying: (1) soluble sodium dichromate and certain soluble dye stuffs will chelate in the presence of acid to yield an insoluble dye stuff firmly bound to the wool fiber, and these colored wools have excellent fastness properties—nylon can be dyed in the same way and (2) on rare occasions dichromate is added to dye stuffs to prevent wool (a

Table 11
INCIDENCE OF CHROMATE SENSITIVITY IN SHOE DERMATITIS

Number of cases of shoe dermatitis	Number positive to chromate	Ref.
35	7	129
165	49	130
64	9	131
213	12	132
42	9	133
42	41	134
25	5	135
100	67	136
25	18	137

reducing agent) from reducing the dye stuff. Lead chromates are very widely used (around 100,000 tons per annum worldwide) for the production of red, yellow, and green paints which exhibit very high opacity and brightness of shade. They are in the insoluble hexavalent form and are thus not relevant as a skin hazard. Chromium oxide green Cr_2O_3, the trivalent form, is widely used in artists' paints and ceramics. Reactions have been described in green pigment tattoos.[140-142] It is possible, however, for trivalent chromium to be oxidized to hexavalent chromium in the tissues.

D. Printing

Chromate dermatitis in the printing industry was first recorded by Parkhurst[82] who reported a case of diffuse eczematous contact dermatitis in a woman working with potassium dichromate in blue print production. The deep etch process most commonly used in the industry today is performed as follows:[143]

The material to be reproduced is photographed and positive transparencies are made. A "grained" metal plate (one with the surface roughened by abrasive treatment) is coated with chromated albumen by the plate maker. The positive photographic films are arranged on the plate in the proper position and sequence. This prepared plate with the film superimposed is then exposed to a carbon arc-light, as is done with ordinary contact prints in photography. Since chromated albumen is sensitive to light, the photographic image is transferred to the prepared plate surface. This image is etched or fixed to the plate with a mixture of gum Arabic, chromic acids and phosphoric acid. The plate is then washed with water which completely removes the chromated albumen from all unfixed areas leaving the image area available for receiving the ink.

The incidence of chromate dermatitis has been variously reported in the printing industry. In Finland, Pirila and Kilpio[88] found that chromate sensitive cases formed 27% (40) of 149 patients with dermatitis in the printing industry which had a total of 18 cases of dermatitis per year per 10,000. Levin et al.[143] estimated that dermatitis affected 5 to 10% of the 30,000 lithographers in the U.S. In their series of 76 cases of dermatitis, 17 gave positive patch tests to potassium dichromate. Contact with chromate in the printing industry occurs with the use of dichromate in the photographic process of making plates. Printers may also be in contact with printing colors containing chromate. Chromate has also been used as an anticorrosive or fungostatic in solutions to keep the press rollers moist.[144] The risk of sensitization to chromate is increased by contact with strong cleansers, degreasing agents, and acids. Recent plate development and a wider choice of anticorrosion agents has rendered the use of dichromate in the printing industry obsolete but it may require some years before it finally disappears from the trade.[144] However, manufacturers may still be unaware of its presence in printing preparations.[145]

E. Glues

Morris[146] described four cases of contact dermatitis to "chrome glue." This glue is made from leather scrap detanned with oxalic acid. The scrap trimmings are washed with water and soaked for weeks or months in milk of lime which causes the fiber to swell. Fatty impurities are converted to insoluble lime soaks which are rinsed and removed by washing. There are two possible sources of chromates: (1) residues from the tanning in the leather and (2) added lime which contributes to the chromium content of the final glue ranging from 0.08 to 17 ppm.[147]

F. Ashes

Fregert[148] believed that dermatitis in a number of patients with chronic chromate dermatitis could be kept going by contact with wood ash. He demonstrated that the chromate content of ashes was of the same order as that present in cement.

G. Foundry Sand

Fregert[149] described three foundry workers with hand dermatitis who, on routine patch testing were found to be sensitive to chromate. Investigation of their working place revealed that they worked with sand containing, among other things, ground chromium magnesite bricks which had been used as refractory material. The foundry sand contained 3 µg chromium trioxide per gram—approximately the same amount as contained in cement.

H. Matches

There is a considerable amount of chromate in a match head, (7.4 to 22.1 mg). This comes from potassium chromate and potassium dichromate added as oxidizers, also as dyes and hardeners in the glue and wood.[150,151] The chromate does not come off on the hands except when wet.[152] When the match is burnt, there is a considerable reduction in the amount of chromate and it changes from hexavalent to trivalent chrome but burnt matches are more easily disintegrated and will liberate a lot more chromate onto the fingers and into the pockets. Fregert[152] found that 33 patients who were allergic to chromate gave positive patch tests with match-heads. He considered that some of these patients improved following avoidance of contact with matches, particularly burnt matches. Fregert further showed that chromate could be demonstrated on the fingers of those who had contact with book matches, even in the dry state. Rudzki and Kozlowska[57] considered that the chromium content of matches was the cause of the high incidence of chromate dermatitis in Poland. The matches contained 5% of dichromate in the ignition composition and 0.64% in the striking composition. They found that 11 patients who kept matches in their pockets and were chromate positive had a rash at the site of contact with the pocket.

I. Wood and Paper Industry

In some wood pulp processes, wood chips are cooked with an alkaline liquor containing sodium hydroxide and sodium sulfide. It is essential to recover these chemicals for use but about 5% is lost and is replaced by sodium sulfate, calcium oxide, and calcium chromate. Samples of sodium sulfate can contain trivalent chromium and also some hexavalent chromium. Fregert et al.[153] described two cases of contact dermatitis in men working in the wood pulp process.

J. Machine Oils

As a general rule, unused, lubricating, or cutting oils contain little or no chromate.[154] Two investigators did find chromate in unused oil (unused bore oil)[155] and a soluble oil.[156] Samitz and Katz[157] could not detect chromate by the diphenyl carbazide test on machinery, work areas, or oil wipe rags. Increased levels of chromium, mostly insoluble, can be found in used cutting oil[158-160] but these are at a very low level. Cases of dermatitis to chromate in mineral oils have been described by Calnan,[156] Holz et al.[155] and Weiler.[161] Ryecroft[162] in a study of soluble oil in MSA lathe machines found it increased from a level of about 0.0 to 0.02 μg/mℓ total filtered chromium to a highest level after 5 weeks of 0.045 μg/mℓ total filtered chromium. None of the 70 operators he studied had a positive patch test to chromate.

In a separate out-patient study at St. John's Hospital, Ryecroft saw only one case of soluble oil dermatitis with a positive chromate patch test. The soluble oil which this man was using contained 0.27 μg/mℓ total chromium after filtration through a 2U membrane. After a study of this patient, it was felt that the chromate sensitivity did not contribute to his dermatitis.

Considering the number of workers in contact with oils of various kinds and the numbers of cases of oil dermatitis, the small numbers of recorded cases of chromate sensitivity in these cases indicate that the chromium in machine oil does not present a significant risk.

K. Timber Preservatives

Copper-chromate-arsenate compounds are used among other compounds to preserve timber. These may be highly fixed (non-water soluble), partially fixed, or unfixed (water soluble), for example, fluor-chrome arsenate.[163] The author has seen two cases of chromate dermatitis due to timber preservatives.

L. Boiler Linings

Considering the large amount of chromium production which goes into refractory linings, it is perhaps unusual that there is only one recorded case of chromate dermatitis due to refractory linings.[164] This may be because most chromate in refractory bricks is trivalent. A case is described of a machinist fitter who worked inside a boiler combustion chamber.[164] Hexavalent chromate was produced from the action of heat and alkaline fuel ash on trivalent chrome ore.

M. Television Workers

In a study of 26 workers[165] involved in making television screens, 18 had chromate dermatitis. In the process, ammonium bichromate is used to produce cross-linking of light sensitive polyvinyl alcohol which enables fluorescent compounds incorporated in the polyvinyl alcohol to adhere to the screen. With improvement in technology, but without changing the process, this has been eliminated as a cause of chromate dermatitis in this work.[166]

N. Magnetic Tapes

Magnetic tapes of a high energy type contain chromium dioxide. Krook et al.[167] described a telegraphist on a ship, who for 4 years, had to daily handle magnetic tapes in a television recorder and he developed dermatitis. Patch testing showed a positive reaction to chromate, cobalt, and nickel.

O. Tire Fitters

Chromate is added to some solutions used to facilitate the reapplying the tire casing to the wheel. Chromate dermatitis due to this solution has been described in tire fitters.[168] Five further cases have since been seen by the author.

P. Chromium Plating

Plating involves the deposition of a layer of chromium on a metal. The metal to be plated is attached to an electrical source and acts as a cathode and with another metal acting as an anode. Usually sulfuric acid is added as a catalyst but heavy metal fluoride can also be used for high speed chromium plating. Chromium trioxide is added, usually in the order of 250 g/ℓ for industrial uses or 400 g/ℓ for especially bright work. There is a high current density and hence a fine spray which makes a very irritant and sensitizing atmosphere. Good ventilation and special protective measures are essential. Royle[169] in a survey of 997 electroplaters found 24.6% had dermatitis. These were not patch tested and many could have been due to other materials. Good prevalence figures for chromate sensitization in the plating industry are not available.

Q. Milk Testers

Potassium dichromate is added to milk as a preservative when stored for fat and protein estimations. Cases of chromate dermatitis to this dichromate have been reported by several authors.[170–172]

R. Food Laboratory

Bang Pedersen[173] described chromate dermatitis in a laboratory technician who was using potassium chromate as an indicator for sodium chloride content in cooked food.

S. Bleaches and Detergents

1. Bleaches

Bleach has been well-recognized as a major source of contact allergy to chromium.[174–176] Lachapelle et al.[177] have indicated that the main reason for addition of chromate is for coloring. Housewives in Europe tend to feel that colorless bleach is not so powerful. It is also used to stabilize the mixture avoiding transformation of hypochlorite into chloride or chlorate and also to clarify the final product, oxidizing and removing impurities. Since 1976, chromate has been removed from Eau de Javelle in France but not in Belgium, Spain, or Italy.[178] Lachapelle et al.[177] showed that liquid bleach in Belgium could contain up to 83 mg chromate per liter. Eau de Javelle has many uses, e.g., domestic-bleaching agent, stain remover, disinfectant, industrial-bleaching of crude textile fibers, chlorination of wool, treatment and purification of water, sterilization of packing, disinfection of materials in slaughter houses, bleaching of oils, soaps, beeswax, corks, and wood pulp. There is, therefore, no doubt that housewives in countries where chromate is added to bleach (and the U.K. and U.S. do not seem to be among those) are at risk from developing chromate dermatitis. Garcia-Perez et al.[178] showed that removing chromate from bleaches reduced the incidence of dermatitis.

2. Detergents

The question of chromate sensitization through contact with detergents is rather more uncertain. Feuerman[77] in Israel reported that on screening 150 housewives with inflammatory skin disease of the hands, 51 were patch test positive to dichromate. Garcia-Perez et al.[178] also quoted a high incidence of chromate sensitivity in hand dermatitis in women in Spain. This source of chromate sensitivity has not been reported from other countries. Small quantities of the order of 0.3 to 6.4 µg/g which appear to derive from the raw material are found in some but not all detergents.[178]

Dr. McMaster, Department of Medicine, Queen's University of Belfast, estimated chromium in nine random samples of detergent and cleaners supplied to her. They varied from 0.4 to 4.4 ppm. When these detergents are diluted in water, the amount of chromium in solution must be very dilute and one must wonder whether it is possible to become sensitized to these amounts of chromium which occur in the U.K. detergents. Schuppli[180] believes the presence of perborates may prevent detection of chromates.

One large manufacturer has provided analytical data for a wide variety of European fabric washing powders as follows:

Number of product packs at each level

	Ni	Cr
Below 1 ppm	71	93
1–2 ppm	17	8
2–5 ppm	16	32
5–10 ppm	31	9
10–15 ppm	18	0
15–20 ppm	0	1

XIII. PREVENTION OF CHROMIUM DERMATITIS

1. Replacement of hexavalent chromate by other compounds. Nitrites and amines are effective antirust agents and can be used instead of chromates. A trivalent chromium plating process is available and is less likely to produce dermatitis and further is a better ecological agent.
2. Reduction of hexavalent to trivalent chromate. Ferrous sulfate added to cement will very effectively reduce hexavalent to trivalent chromate in cement. This is best used in water added to cement on the site rather than at the factory. Except for cost and limitation of the amount of ferrous sulfate available, there is no technical reason why it cannot be added to the factory product. It is available commercially in Sweden (Melstar®).
3. Mechanization. All processes involving chromate should be mechanized in such a way as to minimize handling as far as is reasonably practicable.
4. Protection. Chromate in the atmosphere should be kept at a very low level and if some handling is necessary, protective clothing should be provided though it should be remembered that rubber gloves contain 9.6 to 12.6 ppm chromium and it is possible that this could exacerbate a preexisting dermatitis.[181]
5. Warnings. Good housekeeping and general cleanliness are of the utmost importance. Workers should be warned about the dangers of dermatitis from chromium.
6. Good first aid facilities. Adequate treatment of abrasions and cuts may prevent sensitization.

A. The Effect of Oral Chromium on Dermatitis

The human body contains about 6 mg of chromium. It is thought to be an essential element. Approximately 60 µg average daily intake would be sufficient if availability were high. The average daily intake is between 5 and 115 µg. It is not readily absorbed because any hexavalent chromium in the diet is changed by gastric acidity to the trivalent form which is not absorbed.[182] Various authors have described aggravation of chromate dermatitis from oral chromate. Fregert[183] found that five chromate sensitive patients, to whom he gave 50 µg of chromate, developed severe vesiculation of the palms within 2 hr. Schleiff[184] found that 1 to 10 mg of dichromate by mouth produced exacerbation in 20 chromate sensitive patients. Kaaber and Veien[185] in a double blind study with 2.5 mg chromate and placebo, found that 35% of 31 chromate sensitive patients worsened with the chromate but not with the placebo. The place of oral aggravation of dermatitis by the agent to which the patient is sensitive is still rather a controversial subject.[186] Cases of aggravation to metal prosthesis have been described but nearly all of these have concomitant nickel or chromate allergy which may be the cause of the trouble. It is quite clear that chromate allergy is not a factor in rejection of hip prosthesis (metal head, plastic cup) even in those containing chromium, for example, stainless steel (16 to 18%), and chrome—cobalt (vitallium 26 to 30%).[187-189]

REFERENCES

1. **Cumin, W.**, Remarks on the medicinal properties of madar and the effects of bichromate of potass on the human body, *Edinburgh Med. J.*, 28, 295, 1827.
2. **Dacosta, J. C., Jones, J. F. X., and Rosenberger, R.**, Tanners' ulcer, chrome sores, chrome holes, acid bites, *Ann. Surg.*, 155, 1916.
3. **Blair, J.**, Chrome ulcers. Report of twelve cases, *JAMA*, 90, 1927, 1928.
4. **Legge, T. M.**, *Dermatergoses or Occupational Affections of the Skin*, 4th ed., H. K. Lewis, London, 1934, 447.
5. **Bidstrup, P. L.**, Chromium, alloys, compounds, in 1971. International Labour Office, Geneva, 1971, 294.
6. **Royle, H.**, Toxicity of chromic acid in the chromium plating industry (2), *Environ. Res.*, 10, 141, 1975.
7. **Samitz, M. H. and Epstein, M. D.**, Experimental cutaneous chrome ulcers in guinea pigs, *Arch. Environ. Health*, 5, 463, 1962.
8. **Edmondson, W. F.**, Chrome ulcers of the skin and nasal septum and their relation to patch testing, *J. Invest. Dermatol.*, 17, 17, 1951.
9. **Krishna, G., Mathur, J. S., and Gupta, R. K.**, Health hazard amongst chrome industry workers with special reference to nasal septum perforation, *Indian J. Med. Res.*, 64, 866, 1976.
10. **Jelenko, C.**, Chemicals that "burn," *J. Trauma*, 14, 65, 1974.
11. **Burrows, D. and Cooke, M. A.**, Trivalent chromium plating, *Contact Dermatitis*, 6, 222, 1980.
12. **Rajka, G., Vincze, E., and Csanyi, G.**, Inactivation of eczematogenic industrial chemicals, *Borgyogy. Venerol. Sz.*, 31, 18, 1955.
13. **Dewirtz, A. P.**, Uber Kaliumchromatgeschwure, *Dermatol. Wochenschr.*, 89, 1801, 1929.
14. **Maloof, C. C.**, Use of Edathamil calcium in treatment of chrome ulcers of the skin, *AMA Arch. Ind. Health*, 11, 123, 1955.
15. **Pirozzi, D. J., Gross, P. R., and Samitz, M. H.**, The effect of ascorbic acid on chrome ulcers in guinea pigs, *Arch. Environ. Health*, 7, 178, 1968.
16. **Wall, L. M.**, Chromate eczema and sodium dithionate, *Contact Dermatitis*, in press.
17. **Fregert, S.**, *Manual of Contact Dermatitis*, 2nd ed. Munksgaard, Copenhagen, 1981.
18. **Anderson, F. E.** Cement and oil dermatitis. The part played by chrome sensitivity, *Br. J. Dermatol.*, 72, 108, 1960.
19. **Burrows, D. and Calnan, C. D.**, Cement dermatitis, II, clinical aspects. *Trans. St. John's Hosp. Dermatol. Soc.*, 51, 27, 1965.
20. **Geiser, J. D., Jeanneret, J. P., Delacretaz, J.**, Eczema au ciment et sensibilisation au cobalt, *Dermatologica*, 121, 1, 1960.
21. **Pirila, V.**, On the role of chrome and other trace elements in cement eczema, *Acta Derm. Venereol.*, 34, 136, 1954.
22. **Zelger, J. and Wachter, H.**, Uber die Beziehungen Zwischen Chromat- und Dichromat-Allergie, *Dermatologica*, 132, 45, 1966.
23. **Allenby, C. F. and Goodwin, B. F. J.**, Personal communication, 1981.
24. **Epstein, S.**, Detection of chromate sensitivity. Intradermal versus patch testing, *Ann. Allergy*, 24, 68, 1966.
25. **Metanic. V.**, Prophylake und Behandlung des beginnenden Zementekzems mit Corticosteriod-Salben, *Berufs-Dermatosen*, 13, 288, 1965.
26. **Spier, H. W. and Natzel, R.**, Chromat-allergie und Zementekzem, *Hautarzt*, 4, 63, 1953.
27. **Miescher, G., Amrein, H. P., and Leder, M.**, Chromatuberempfindlichkeit und Zementekzem, *Dermatologica*, 110, 266, 1955.

28. **Skog, E. and Wahlberg, J. E.**, Patch testing with potassium dichromate in different vehicles, *Arch. Dermatol.*, 99, 697, 1969.
29. **Rudzki, E., Zakrzewski, Z., Prokopczyk, G., and Kozlowska, A.**, Application of emulsifiers for the patch test, *Dermatologica*, 153, 333, 1976.
30. **Engelhardt, W. E. and Mayer, R. L.**, Ueber Chromekzemeim graphischen Gewerbe, *Arch. Gewerbepathol. Gewerbehyg.*, 2, 140, 1931.
31. **Fregert, S. and Rorsman, H.**, Allergy to trivalent chromium, *Arch. Dermatol.*, 90, 4, 1964.
32. **Fregert, S. and Rorsman, H.**, Allergic reactions to trivalent chromium compounds, *Arch. Dermatol.*, 93, 711, 1966.
33. **Forstrom, L., Pirila, V., and Huju, P.**, Rehabilitation of workers with cement eczema due to hypersensitivity to bichromate, *Scand. J. Rehabil. Med.*, 1, 95, 1969.
34. **Calnan, C. D.**, Personal communication, 1981.
35. **Rudzki, E., Zakrzewski, G., Prokopczyk, G., and Kozlowska, A.**, Contact sensitivity to trivalent chromium compounds, *Dermatol. Beruf. Umwelt*, 26, 83, 1978.
36. **Cronin, E.**, *Contact Dermatitis*, Churchill Livingstone, London, 1980, 4.
37. **Nielson, Von C. and Klaschka, F.**, Teststudien an der Mundschleimhaut bei Ekzemallergikern, *Dtsch. Zahn Mund Kieferheilkd.*, 57, 201, 1971.
38. **Amrein, H. P. and Miescher, G.**, Zur Atiologie des Zementekzems; Reihenuntersuchungen uber die Haufigkeit einer Chromatsensibilisierung im Maurerewerbe, *Z. Unfallmed. Berufskr.*, 48, 140, 1955.
39. **Hunziker, N. and Musso, E.**, A propos de l'eczema au ciment, *Dermatologica*, 121, 204, 1960.
40. **Walsh, E. N.**, Chromate hazards in industry, *JAMA*, 153, 1305, 1953.
41. **Bett, D. C. G.**, The potassium dichromate patch test, *Trans. St. John's Hosp. Dermatol. Soc.*, 40, 40, 1958.
42. **Jaeger, H. and Pelloni, E.**, Tests epicutanes aux bichromates, positifs dans l'eczema au ciment, *Dermatologica*, 100, 207, 1950.
43. **Stevenson, C.**, Personal communication, 1960.
44. **Cronin, E.**, clinical prediction of patch test results, *Trans. St. John's Hosp. Dermatol. Soc.*, 58, 153, 1972.
45. **Thormann, J., Jespersen, N. B., and Joensen, H. D.**, Persistence of contact allergy to chromium, *Contact Dermatitis*, 5, 261, 1979.
46. **Burry, J. N., Kirk, J., Reid, J. G., and Turner, T.**, Environmental dermatitis; patch test in 1,000 cases of allergic contact dermatitis, *Med. J. Aust.*, 2, 681, 1973.
47. **Lachapelle, J. M. and Tennstedt, D.**, Epidemiological survey of occupational contact dermatitis of the hands in Belgium, *Contact Dermatitis*, 5, 244, 1979.
48. **Moriearty, P. L., Pereira, C., and Guimaraes, N. A.**, Contact dermatitis in Salvador, Brazil, *Contact Dermatitis*, 4, 185, 1978.
49. **Epstein, E., Rees, W. J., and Maibach, H. I.**, Recent experiences with routine patch test screening, *Arch. Dermatol.*, 98, 18, 1968.
50. **Hammershoy, O.**, Standard patch test results in 3,225 consecutive Danish patients from 1973 to 1977, *Contact Dermatitis*, 6, 263, 1980.
51. **Fregert, S., Hjorth, N., Magnusson, B., Bandmann, H.-J., Calnan, C. D., Cronin, E., Malten, K., Meneghini, C. L., Pirila, V., and Wilkinson, D. S.**, Epidemiology of contact dermatitis, *Trans. St. John's Hosp. Dermatol. Soc.*, 55, 17, 1969.
52. **Burrows, D.**, Chromium and the skin, *Br. J. Dermatol.* 99, 587, 1978.
53. **Kanan, M. W.**, Cement dermatitis and atmospheric parameters in Kuwait, *Br. J. Dermatol.*, 86, 155, 1972.
54. **Baer, R. L., Ramsey, D. L., and Biondi, E.**, The most common contact allergens, *Arch. Dermatol.*, 74, 108, 1973.
55. North American Contact Dermatitis Group, Epidemiology of contact dermatitis in North America, 1972, *Arch. Dermatol.*, 108, 537, 1973.
56. **Rudzki, E. and Kleniewska, D.**, The epidemiology of contact dermatitis in Poland, *Br. J. Dermatol.*, 83, 543, 1970.
57. **Rudzki, E. and Kozlowska, A.**, Causes of chromate dermatitis in Poland, *Contact Dermatitis*, 6, 191, 1980.
58. **Magnusson, B., Blohm, S-G., Fregert, S., Hjorth, N., Hovding, G., Pirila, V., and Skog, E.**, Routine patch testing IV, *Acta Derm.-Venereol.*, 48, 110, 1968.
59. **Husain, S. L.**, Contact dermatitis in the West of Scotland, *Contact Dermatitis*, 3, 327, 1977.
60. **Camarasa, J. M. G., Alomar, A., Botella, R., Conde, L., Perez, A. G., Orbaneja, J. G., Grimalt, F., Marquina, A., Pascual, J. M., Ocana, J., and Romaguera, C.**, First epidemiological study of contact dermatitis in Spain—1977, Acta Derm.-Venereol., 59(Suppl. 85), 33, 1979.

61. **Marcussen, P. V.,** Variations in the incidence of contact hypersensitivities, *Trans. St. John's Hosp. Dermatol. Soc.,* 48, 40, 1962.
62. **Baer, R. L., Lipkin, G., Kanof, N. B., and Biondi, E.,** Changing patterns of sensitivity to common contact allergens, *Arch. Dermatol.,* 89, 3, 1964.
63. **Hannuksela, M. and Pirila, V.,** Quoted in Rudzki et al., 60, Personal communication, 1976.
64. **Rudzki, E., Zakrzewski, Z., Prokopczyk, G., and Kozlowska, A.,** Patch tests with potassium dichromate removed after 24 and 48 hours, *Contact Dermatitis,* 2, 309, 1976.
65. **Magnusson, B., Fregert, S., Hjorth, N., Hovding, G., Pirila, V., and Skog. E.,** Routine patch testing. V, *Acta Derm.-Venereol.,* 49, 556, 1969.
66. **Kanan, M. W.,** Contact dermatitis in Kuwait, *J. Kwt. Med. Assoc.,* 3, 129, 1969.
67. **Cronin, E.,** *Contact Dermatitis,* Churchill Livingstone, London, 1980, 292.
68. **Whaba, A. and Cohen T.,** Chrome sensitivity in Israel, *Contact Dermatitis,* 5, 101, 1979.
69. **Cronin, E.,** *Contact Dermatitis,* Churchill Livingstone, London, 1980, 294.
70. **Shanon, J.,** Pseudo-atopic dermatitis, *Dermatologica,* 131, 176, 1965.
71. **Engel, H. O. and Calnan, C. D.,** Chromate dermatitis from paint, *Br. J. Ind. Med.,* 20, 192, 1963.
72. **Hall, A. F.,** Occupational contact dermatitis among aircraft workers, *JAMA,* 125, 179, 1944.
73. **Bernhardt, H. J.,** Chromate dermatitis. Its natural history and treatment, *Arch. Dermatol.,* 76, 13, 1957.
74. **Mitchell, J. C.,** The angry back syndrome; eczema creates eczema, *Contact Dermatitis,* 1, 193, 1975.
75. **Frain-Bell, W.,** What is this thing called light?, *Clin. Exp. Dermatol.,* 4, 1, 1979.
76. **Wahlberg, J. E. and Wennersten, G.,** Light sensitivity and chromium dermatitis, *Br. J. Dermatol.,* 97, 411, 1977.
77. **Feuerman, E. J.,** The relevance of sensitivity to chromate in woman, *Br. J. Dermatol.,* 82, 205, 1970.
78. **Breit, R. and Turk, R. B. M.,** The medical and social fate of the dichromate allergic patient, *Br. J. Dermatol.,* 94, 349, 1976.
79. **Burrows, D.,** Prognosis in industrial dermatitis, *Br. J. Dermatol.,* 87, 145, 1972.
80. **Fregert, S.,** Occupational dermatitis in a 10 year material, *contact Dermatitis,* 1, 96, 1975.
81. **Purschel, W. and Furst, G.,** Berufsbedingtes Kontaktekzem, Katasnesen und Rehabilitation, *Berufs-Dermatosen,* 20, 174, 1972.
82. **Parkhurst, H. J.,** Dermatosis industralis in a blue print worker due to chromium compounds, *Arch. Dermatol. Syphilol.,* 12, 253, 1925.
83. **Smith, A. R.,** Chrome poisoning with manifestations of sensitization. Report of a case, *JAMA,* 97, 95, 1931.
84. **Calnan, C. D.,** Cement dermatitis, *J. Occup. Med.,* 2, 15, 1960.
85. **O'Donavan, W. J.,** Linedermatitis, *Lancet,* 1, 599, 1925.
86. **Bonnevie, P.,** *Aetologie und Pathogenese der Ekzemkrankheiten,* Busck, Copenhagen-Leipzig, 1939.
87. **Stauffer, H.,** Die Ekzemproben (Methodik und Ergebnisse), *Arch. Dermatol. Syphilol.,* 162, 517, 1939.
88. **Pirila, V. and Kilpio, O.,** On dermatoses caused by bichromates, *Acta Derm.-Venereol.* 29, 550, 1949.
89. **Cronin, E.,** Chromate dermatitis in men, *Br. J. Dermatol.,* 85, 95, 1971.
90. **Cronin, E.,** *Contact Dermatitis,* Churchill Livingstone, London, 1980. 296.
91. **Rabito, C. and Peserico, A.,** Sull'etiologia della sensibilizzaziono nell'eczema da cemento, *Minerva Dermatol.,* 108, 287, 1973.
92. **Delacretaz, J. and Geiser, J. D.,** Nouvelles recherches sur les facteurs allergiques dans l'eczema au ciment, *Symposium Dermatologorum Prague,* 1960, 245.
93. **Engebrigsten, J. K.,** Some investigations on hypersensitiveness to bichromate in cement workers, *Acta Derm.-Venereol.,* 32, 462, 1952.
94. **Frey, E.,** Beitrag zur Bewertung der Jaegerschen Kalibicromatprobe auf Zement, *Dermatologica,* 105, 244, 1952.
95. **Gomez, L. C-S. and Urcuyo, J. F. G.,** Sensibilidad a los componentes de la goma en obreros de la construccion, (revision de 100 casos), *Acta Dermo-Sifiliograficas,* 67, 297, 1976.
96. **Hilt, M. G.** La dermite du chrome hexavalent dans le cadre des dermites eczemateuses par sensibilisation aux metaux, *Dermatologica,* 109, 143, 1954.
97. **Hovding, G.,** Cement eczema and chromium allergy. An epidemiological investigation, Ph.D. thesis, University of Bergen, Norway, 1970, 101.
98. **Lejhancova, G. and Wolf, J.,** Die Bedeutung des Kaliumbichromattestes beim Alkaliekzem, *Dermatol. Wochenschr.,* 132, 1273, 1955.
99. **Perone, V. B., Moffitt, A. E., Possick, P. A., Key, M. M., Danzinger, S. J., and Gellin, G. A.,** The chromium, cobalt and nickel contents of American cement and their relationship to cement dermatitis, *Am. Ind. Hyg. Assoc. J.,* 35, 301, 1974.

100. **Pirila, V. and Kajanne, H.,** Sensitization to cobalt and nickel in cement eczema, *Acta Derm.-Venereol.,* 45, 9, 1965.
101. **Johnston, A. J. M. and Calnan, C. D.,** Cement Dermatitis. I. Clinical Aspects, *Trans. St. John's Hosp. Dermatol. Soc.,* 41, 11, 1958.
102. **Fregert. S. and Gruvberger, B.,** Chemical properties of cement, *Berufs-Dermatosen,* 20, 238, 1972.
103. **Huriez, C.,** Occupational skin diseases, *Acta Derm.-Venereol.,* Proc. 11th Int. Congr. Dermatol., 2, 245, 1957.
104. **Huriez, C., Martin, P., and Planque, Mme.,** Dermites des Cimentiers, *Ann. Dermatol. Syphiligr.,* 96, 375, 1965.
105. **Geiser, J. D. and Girard, A.,** Remarques sur les cas d'eczema au ciment observes a la Clinique de dermato-venereologie de Lausanne de 1947 a 1961, *Dermatologica,* 131, 93, 1965.
106. **Meneghini, C. L. and Petruzzelis, V.,** Incidence of dermatitis in cement workers, *Contact Dermatitis Newsl.,* 3, 55, 1968.
107. **Wahlberg, J. E.,** Health screening for occupational skin diseases in building workers, *Berufs-Dermatosen,* 17, 184, 1969.
108. **Varigos, G. A. and Dunt, D. R.,** Occupational dermatitis. An epidemiological study in the rubber and cement industries, *Contact Dermatitis,* 7, 105, 1981.
109. **Adams, R. M., Fregert, S., Gruvberger, B., and Maibach, H. I.,** Water solubility of zinc chromate primer paints used as anti-rust agents, *Contact Dermatitis,* 2, 357, 1976.
110. **Bang Pedersen, N. and Fregert, S.,** Primer on a leg prosthesis as a source of chromate, *Contact Dermatitis Newsl.,* 8, 191, 1970.
111. **Fregert, S., Gruvberger, B., and Heijer, A.,** Chromium dermatitis from galvanised sheets, *Berufs-Dermatosen,* 18, 254, 1970.
112. **Newhouse, M. L.,** A cause of chromate dermatitis among assemblers in an automobile factory, *Br. J. Ind. Med.,* 20, 199, 1963.
113. **Winston, J. R. and Walsh, E. N.,** Chromate dermatitis in railroad employees working with diesel locomotives, *JAMA,* 147, 1133, 1951.
114. **Calnan, C. D. and Harman, R. R. M.,** Studies in contact dermatitis, XIII, Diesel Coolant Chromate Dermatitis, *Trans. St. John's Hosp. Dermatol. Soc.,* 46, 13, 1961.
115. **Plunkett, R.,** Diesel engine dermatitis control, *Minn. Med.,* 37, 336, 1954.
116. **Haeberlin, J. B.,** Dermatitis in the railroad industry, *Ind. Med.,* 24, 255, 1955.
117. **Guy W. B.,** Dermatologic problems in the railroad industry resulting from conversion to diesel power, *Arch. Dermatol. Syphilol.,* 70, 289, 1954.
118. **Marsland, T.,** Water, fuel and oil problems, *Diesel Railway Traction,* 301, 1957.
119. **Calnan, C. D.,** Chromate in coolant water of gramophone record presses, *Contact Dermatitis,* 4, 346, 1978.
120. **Ros, A. M. and Bang Pedersen, N.,** Chromate in a defatting solvent, *Contact Dermatitis,* 3, 105, 1977.
121. **Wilson, H. T. H.,** Chrome dermatitis in a brewery, *Contact Dermatitis Newsl.,* 10, 228, 1971.
122. **Fregert, S. and Ovrum, P.,** Chromate in welding fumes with special reference to contact dermatitis, *Acta Derm.-Venereol.,* 43, 119, 1963.
123. **Shelley, W. B.,** Chromium in welding fumes as cause of eczematous hand eruption, *JAMA,* 189, 772, 1964.
124. **Thrysin, E., Gerhardsson, G., and Forssman, S.,** Fumes and gases in arc welding, *AMA Arch. Ind. Hyg. Occup. Med.,* 6, 381, 1952.
125. **Sykes, R. L.,** Director, British Leather Manufacturing Association, Personal communication, 1981.
126. **Roddy, W. T. and Lollar, R. M.** Resistence of white leather to breakdown by perspiration, *J. Am. Leather Chem. Assoc.,* 50, 180, 1955.
127. **Samitz, M. H. and Gross, S.,** Extraction by sweat of chromium from chrome tanned leathers, *J. Occup. Med.,* 2, 12, 1960.
128. **Morris, G. E.,** "Chrome" dermatitis, *Arch. Dermatol.,* 78, 612, 1958.
129. **Adams, R. M.,** Shoe dermatitis, *Calif. Med.,* 117, 12, 1972.
130. **Angelini, G., Vena, G. A., and Meneghini, C. L.,** Shoe contact dermatitis, *Contact Dermatitis,* 6, 279, 1980.
131. **Calnan, C. D. and Sarkany, I.,** Studies in contact dermatitis. IX, Shoe Dermatitis, *Trans. St. John's Hosp. Dermatol. Soc.,* 43, 8, 1959,
132. **Cronin, E.,** Shoe dermatitis, *Br. J. Dermatol.,* 78, 617, 1966.
133. **Dahl, M. V.,** Allergic contact dermatitis from footwear. Shoe dermatitis in 42 patients, *Minn. Med.,* 58, 871, 1975.
134. **Hinson, T. C.,** Chrome leather dermatitis in Singapore, *Contact Dermatitis Newsl.,* 6, 120, 1969.
135. **Jordan, W. P.,** Clothing and shoe dermatitis. Recognition and management, *Postgrad. Med.,* 52, 143, 1972.

136. **Scutt, R. W. B.**, Chrome sensitivity associated with tropical footwear in the Royal Navy, *Br. J. Dermatol.*, 78, 337, 1966.
137. **Varelzides, A., Katsambas, A., Georgala, S., and Capetanakis, J.**, Shoe dermatitis in Greece, *Dermatologica*, 149, 236, 1974.
138. **Fregert, S. and Gruvberger, B.**, Chromium in industrial gloves, *Contact Dermatitis*, 5, 189, 1979.
139. **Cronin, E.**, *Contact Dermatitis*, Churchill Livingstone, London, 1980, 305.
140. **Bjornberg, A.**, Allergic reactions to chrome in green tattoo markings, *Acta Derm.-Venereol.*, 39, 23, 1959.
141. **Loewenthal, L. J. A.**, Reactions in green tattoos, *Arch. Dermatol.*, 82, 237, 1960.
142. **Cairns, R. J. and Calnan, C. D.** Green tattoo reactions associated with cement dermatitis, *Br. J. Dermatol.*, 74, 288, 1962.
143. **Levin, H. M., Brunner, M. J., and Rattner, H.**, Lithographer's dermatitis, *JAMA*, 169, 566, 1959.
144. **Adams, R. M.**, Allergen replacement in industry, *Cutis*, 20, 511, 1977.
145. **Spruit, D. and Malten, K. E.**, Occupational cobalt and chromium dermatitis in an offset printing factory, *Dermatologica*, 151, 34, 1975.
146. **Morris, G. E.**, Chromate dermatitis from chrome glue and other aspects of the chrome problem, *AMA Arch. Ind. Health*, 11, 368, 1955.
147. **Weiler, K. J. and Russel, H. A.**, Das Chromekzem durch Glutinleim and Kalk, *Berufs-Dermatosen*, 19, 292, 1971.
148. **Fregert, S.**, The chromium content of fuel ashes with reference to contact dermatitis, *Acta Derm.-Venereol.*, 42, 476, 1962.
149. **Fregert, S.**, Contact dermatitis due to chromate in foundry sand, *Acta Derm.-Venereol.*, 43, 477, 1963.
150. **Schwartz, L., Tulipan, L., and Birmingham, D. J.**, *Occupational Diseases of the Skin*, 3rd ed., Lea & Febiger, Philadelphia, 1957, 873.
151. **Udy, J.**, *Chromium*, Vol. 1, Reinhold, New York, 1956, 389.
152. **Fregert, S.**, Chromate eczema and matches, *Acta Derm.-Venereol.*, 41, 433, 1961.
153. **Fregert, S., Gruvberger, B., and Heijer, A.**, Sensitization to chromium and cobalt in processing of sulphate pulp, *Acta Derm.-Venereol.*, 52, 221, 1972.
154. **Wahlberg, J. E., Linstedt, G., and Einarsson, O.**, Chromium, cobalt and nickel in Swedish cement, detergents mould and cutting oil, *Berufs-Dermatosen*, 25, 220, 1977.
155. **Holz, H., Mappes, R., and Weidmann, G.**, Chromatallergie bei Bohrolekzem, *Berufs-Dermatosen*, 9, 113, 1961.
156. **Calnan, C. D.**, chromate dermatitis from soluble oil, *Contact Dermatitis*, 4, 378, 1978.
157. **Samitz, M. H. and Katz, S. A.**, Skin hazards from nickel and chromium salts in association with cutting oil operations, *Contact Dermatitis*, 1, 158, 1975.
158. **Russell, H. A. and Weiler, K.-J.**, Uber das Vorkommen von Chrom in gebrauchten Mineralolen, *Berufs-Dermatosen*, 19, 23, 1971.
159. **Einarsson, O., Kylin, B., Linstedt, G., and Wahlberg, J. E.**, Chromium, cobalt and nickel in used cutting fluids, *Contact Dermatitis*, 1, 182, 1975.
160. **Oleffe, J., Roosels, D., Vanderkeel, J. and Groetenbriel, C.**, Presence du chrome dans l'environment de travail, *Berufs-Dermatosen*, 19, 57, 1971.
161. **Weiler, K.-J.**, Rezidivierendes Chromkontaktekzem durch Umgang mit Chromstahl, *Berufs-Dermatosen*, 17, 316, 1969.
162. **Rycroft, R.**, for Doctorate of Medicine Thesis, Cambridge University.
163. **Wilkinson, D. S.**, Timber preservatives, *Contact Dermatitis*, 5, 278, 1979.
164. **Rycroft, R. J. G. and Calnan, C. D.**, Chromate dermatitis from a boiler lining, *Contact Dermatitis*, 3, 198, 1977.
165. **Stevenson, C. J.**, Fluorescence as a clue to contamination in TV workers, *Contact Dermatitis*, 1, 242, 1975.
166. **Stevenson, C. J. and Morgan, P. R.**, Investigation and prevention of chromate dermatitis in colour television manufacture, *J. Soc. Occup. Med.*, in press.
167. **Krook, G., Fregert, S., and Gruvberger, B.**, Chromate and cobalt eczema due to magnetic tapes, *Contact Dermatitis*, 3, 60, 1977.
168. **Burrows, D.**, Chromium dermatitis in a tyre fitter, *Contact Dermatitis*, 7, 55, 1981.
169. **Royle, H.**, Toxicity of chromic acid in the chromium plating industry (2), *Environ. Res.*, 10, 141, 1975.
170. **Huriez, C., Martin, P., and Lefebvre, M.**, Sensitivity to dichromate in a milk analysis laboratory, *Contact Dermatitis*, 1, 247, 1975.
171. **Rogers, S. and Burrows, D.**, Contact dermatitis to chrome in milk testers, *Contact Dermatitis*, 1, 387, 1975.

172. **Rudszki, E. and Czerwinska-Dihnz, I.**, Sensitivity to dichromate in milk testers, *Contact Dermatitis*, 3, 107, 1977.
173. **Bang Pedersen, N.**, Chromate in a food laboratory, *Contact Dermatitis*, 3, 105, 1977.
174. **Rabeau, H. and Ukrainczyk, F.**, Dermites des "Blanchisseuses," role du chrome et du chlore en France, *Ann. Dermatol. Syphiligr.*, 10, 656, 1939.
175. **Sidi, E. and Hincky, M.**, Problemes d'actualite concernani les dermatoses professionnelles, *Rev. Fr. D'Allerg.*, 5, 198, 1965.
176. **Hilt, S.**, La dermite du chrome hexavlanet dans le cadre des Dermites Eczemateuses for sensibilisation adx Metzux, *Dermatologica*, 104, 143, 1954.
177. **Lachapelle, J. M., Cauwerys, R., Tennstedt, D., Andanson, J., Benezra, C., Chabeau, G., Ducombs, G., Foussereau, J., Lacroix, M., and Martin, P.**, Eau de Javel and prevention of chromate allergy in France, *Contact Dermatitis*, 6, 107, 1980.
178. **Garcia-Perez, A., Martin-Pascual, A., and Sanchez-Misiego, A.**, Chrome content in bleaches and detergents: its relationship to hand dermatitis in women, *Acta Dermatol.*, 53, 353, 1973.
179. **Nater, J. P.**, Possible causes of chromate eczema, *Dermatologica*, 126, 160, 1963.
180. **Schuppli, R.**, Synthetische waschnuttel und metallionen, *Dermatologica*, 135, 403, 1967.
181. **Conde-Salazar, L., Castano, A., and Vincente, M. A.**, Determination of chrome in rubber gloves, *Contact Dermatitis*, 6, 237, 1980.
182. National Research Council Committee on biological effects of atmospheric pollutants. Medical and biological effects of environmental pollutants, chromium, National Academy of Sciences, Washington, D.C. 1974, 28.
183. **Fregert, S.**, Sensitization to Hexa- and Trivalent Chromium, Proc. Congr. Hung. Derm. Soc., April 1965, 50.
184. **Schleiff, P.**, Provokation des Chromatekzems zu Testzwecken durch interne Chromzufuhr, *Hautarzt*, 19, 209, 1968.
185. **Kaaber, K. and Veien, N. K.**, The significance of chromate ingestion in patients allergic to chromate, *Acta Derm.-Venereol.*, 57, 321, 1977.
186. **Burrows, D., Creswell, S., and Merrett, J. D.**, Nickel, hands and hip, *Br. J. Dermatol.*, 105, 437, 1981.
187. **Carlsson, A. S., Magnusson, B., and Moller, H.**, Metal sensitivity in patients with metal-to-plastic total hip arthroplasties, *Acta Orthop. Scand.*, 51, 57, 1980.
188. **Rooker, G. D. and Wilkinson, J. D.**, Metal sensitivity in patients undergoing hip replacement, *J. Bone Jt. Surg.*, 62B, 502, 1980.
189. **Deutman, R., Mulder, T. J., Brian, R., and Nater, J. P.**, Metal sensitivity before and after total hip arthroplasty, *J. Bone Jt. Surg.*, 59, 862, 1977.

INDEX

A

Absorption, 3
  atomic, 9
  hexavalent chromium, 66
  internal, 3
Accessory cells, 82
Accuracy, 8
Acetate, 56, 142
Acid buffering, 53
Activation
  analysis of, 6
  enhanced, 88
  suppressor cells, 88
Adenocarcinomas, 24
Adenomas
  alveologenic, 24
  pulmonary, 45
Air
  environmental, 14
  monitoring of, 33
Airway obstruction, 32
Albumin, 69
  bovine serum, 74
  chromium, 69
  human serum, 69
Alkaline buffering, 53
Allergic contact dermatitis, 52—55
Allergy to chromium, 57, 148
  subliminal, 110, 111
ALS, see Antilymphocyte serum
Alveologenic adenomas, 24
Amino acids, 69
Ammonium bichromate, 146, 155
Analytical techniques, 5—10
Anaphylaxis, 117
Animal fats, 3
Animal models, 23—24, 61
Animals, see also specific animals
  experimental studies in, 23—25
  induction of tumors in, 45—49
Anion, 68
Antibodies, 58
  circulating, 70
Antibody-mediated skin reactions, 58
Antichrome powder, 139
Anticorrosives, 153
Antigen-reactive cells, 103
Antigens
  immune response gene associated (Ia), 84
  synthetic, 84, 86, 87
Antihistamines, 119
Anit-lymph node permeability factor, 114
Antilymphocyte serum (ALS), 96, 106, 107
  cyclophosphamide and, 96—98, 107
  heterologous, 94
  prednisolone and, 98
Antipolymorphonuclear serum, 114, 115
Antirust agents, 149—151

Antithymocyte serum (AThS), 94
Arsenic, 87, 119
Arthus reaction, 111, 114, 116, 118, 120
Ascorbic acid, 139, 140
Ashes, 154
Associated light sensitivity, 146
Asthma, 32, 34—35
AThS, see Antithymocyte serum
Atmospheric concentrations, 17
Atomic absorption spectrophotometry, 5, 9
Atomic fluorescence, 5

B

Barium chromate, 24
Basophils, 116
Beryllium fluoride, 84—86, 89, 92, 109, 111, 121
Beryllium lactate, 92, 104, 109, 121
Bichromate-producing industry, 49
Binding capacity, 73
  hexavalent chromium, 68
  proteins, 67—69, 82
  trivalent chromium, 68
Binding of chromium, 4
Biological aspects, 2—5
Biological dose response to trace elements, 2
Bleaches, 156
Blockade of receptors, 100, 104
Boiler linings, 155
Bovine serum albumin (BSA), 74
Bradykinine, 122
Brewery, 151
Bronchial tumors, 24
Bronchitis, 34
Bronchogenic carcinoma, 4, 35, 45
Bronchoscopy, 37
Bronchospasm, 32, 34
Brown sugar, 3
BSA, see Bovine serum albumin
Buffering, 53
Butter, 3

C

Calcium chromate, 24, 25, 36, 46—48
Calcium disodium salt of ethylemidiamine tetra acetic acid, 139
Calling effect, 104
Cancer, see also Carcinoma; Sarcoma; specific types of cancer
  gastrointestinal tract, 15, 20—22, 25
  lung, see Lung cancer
  nasal, 19, 34
  occupational, 18
Carcinogenic hazards, 14—15, 18
Carcinoma, see also Cancer
  bronchogenic, 4, 35, 45
  squamous, 45, 47

Carriers
    protein, 67, 69
    specificity of, 59
Case-referent studies, 17
Cation, 68
Cell-mediated skin reactions, 58
Cells, see also specific types of cells
    accessory, 82
    antigen-reactive, 103
    dendritic, 59
    effector, 100, 103, 104
    Langerhans', 59, 63, 67, 70, 84, 109, 117, 118
    memory, 103, 104
    mononuclear, 119
    polynuclear, 119
    squamous, 45
    stimulator, 74, 87, 118
    suppressor, see Suppressor cells
    T, 58, 84, 87
Cement, 53, 65, 147—149
Cement dermatitis, 149
Cement eczema, 53, 56, 148
Chemical irritation, 32, 34
Chemotactic factor for eosinophils, 122
Chloride, 56, 142
Chromic acid, 22, 26, 48, 139
Chromic chloride, 142
Chromium albumin complex, 69
Chromium chloride, 69, 81, 104
Chromium chloride-GPS conjugate, 76
Chromium-cystin conjugate, 69
Chromium-modified macrophages, 81
Chromium oxide, 153
Chromium sulfate, 92, 111
Chromium trichloride, 142
Chromium trioxide, 15, 22, 24—26, 150
Chromyl chloride, 48
Chronic bronchitis, 34
Circulating antibodies, 70
Circulating immune complexes, 121
Clonal deletion, 88, 103
Clonal selection theory, 87, 101
Cobalt, 69
    reactions to, 55
Cohort studies, 17
Colorimetry, 64
Common determinant, 81, 83
Con A, see Concanavalin A
Concanavalin A (Con A), 74
Concentrations
    atmospheric, 17
    maximum allowable, 19
    threshold, 55, 56
Conjugates
    chromium chloride-GPS, 76
    chromium-cystin, 69
    dinitrophenyl, 84
    trivalent chromium, 69
Contact dermatitis, 146—145
    allergic, 52—55
Contact sensitivity, 72
    desensitization in, 87—107

DNCB, see Dinitrochlorobenzene
    experimental aspects of, 61—64
    genetic factors in, 84—87
    immunological mechanisms of, 57—61
    in-vitro reactions associated with, 73—82
    metal, 84—87
    secondary phenomena during, 107—122
    tolerance in, 87—107
Contamination of environment, 14
Control of sources of exposure, 33
Coolant antirust agents, 149—151
Copolymers, 84
Covalent bonds with proteins, 69
Cross-reactions, 71
Cross-sensitization, 71
Cutaneous anaphylaxis, 117
CY, see Cyclophosphamide
Cyclophosphamide (CY), 94, 96, 106
    antilymphocyte serum and, 96—98, 107
    prednisolone and, 98
Cysteine, 69
Cystine, 69
Cytology, 37

D

Daily intake, 3
Defatting solvents, 151
Dendritic cells, 59
Dermatitis, 4, 32, 57, 149, 157—158
    allergic contact, 52—55
    cement, 149
    contact, see Contact dermatitis
    oral chromium and, 157—158
    patterns of, 144—146
    shoe, 152, 153
Desensitization, 88—98, 100, 103
    contact sensitivity and, 87—107
    mechanisms of, 88, 101—107
    permanent, 103, 104, 106, 112
    procedures for, 88
Detection limit, 8
Detergents, 156—157
Determinants, 81, 83
Development time, 16
Diabetics, 3
Diagnosis of lung cancer, 36—40
Diet, 14
Diffusion precipitation, 117
Dilution of materials used in patch tests, 141
Dimethylsulfoxide (DMSO), 62
Dinitrobenzenesulfonic acid, 108
Dinitrochlorobenzene (DNCB), 75, 88, 92, 99, 103, 104, 108, 119—121
Dinitrophenyl conjugates, 84
Direct hemolysis, 117
Direct MMT, 101
Disease treatment with chromium, 15
Dissociation constants of chromium compounds, 83
Distribution
    chromium compounds, 67

spatial, 9
DMSO, see Dimethylsulfoxide
DNA synthesis, 60, 73
$DNBSO_3$, 104
DNCB, see Dinitrochlorobenzene
Dose
   definition of, 17
   dependence on, 91
   estimation of, 17, 18
   minimum initiation, 15, 17
   sufficient, 15
Dose-response relationships, 17, 22, 25—26
   biological, 2
Double diffusion precipitation in gel, 117
Double-shot phenomenon, 92, 103, 104, 106, 112, 119
Draining lymph node, 60
Drinking water, 3, 14
Drug eruptions, 109
Duration of exposure, 17
Dust, 24
Dyes, 152

E

Early diagnosis of lung cancer, 36—40
Earth's crust, 2, 3
Eczema, 57
   cement, 53, 56, 148
   chromium, 54
   leather, 56
Eczematogenic exposures to chromium compounds, 54
EDTA, 139
Effector cells, 100, 103, 104
Electron microprobes (EMP), 7
Elementary chromium, 83
Elementation composition, 9
Eliciting phase, 60
EMP, see Electron microprobes
Emulsifiers, 141
Enhanced activation, 88
Environmental air, 14
Environmental contamination, 14
Environmental exposure, 4—5
Eosinophils, 122
Epicutaneous application, 89
Epicutaneous challenges, 99
Epicutaneous tests, 141
Epidemiology, 15—23, 35
   of lung cancer, 35—40
Epidermal challenge, 104
Ethylemidiamine tetra acetic acid, 139
Excretion of chromium, 4, 37
Experimental animals, 23—25
   induction of tumors in, 45—49
Experimental aspects of contact sensitivity to chromium, 61—64
Exposure
   characterization of, 26
   control of sources of, 33
   duration of, 17
   eczematogenic, 54
   environmental, 4—5
   inhalation, 24
   normal, 2
   occupational, 4, 14
   time of, 15
Exudate, 76, 78, 104

F

Fats, 3
FCA, see Freund's complete adjuvant
Feeding, 89, 98
Ferrochromium, 14, 20
Ferrous sulfate, 157
Fibrosarcoma of thigh muscle, 46
First-shot, 104, 105
Fixed drug eruptions, 109
Flame emission spectrometry, 9
Flameless atomic absorption spectrophotometry, 5
Flare-up reactions, 108, 115, 119
   nonspecific, 120
   secondary, 109—120
   spontaneous, 109, 120
Fluorescence
   atomic, 5
   X-ray, 8
Food laboratory, 156
Foods, 3, 15
Foundry sand, 154
Freunds's complete adjuvant (FCA), 62
   potassium dichromate in, 62, 75
Freund's incomplete adjuvant, 99
Fumes from welding, 22, 25
Fungostatics, 153

G

Galvanizing, 150
Gastrointestinal tract cancer, 15, 20—22, 25
Generalized rash, 120—122
Genetic factors in contact sensitivity to metals, 84—87
Globulin, 69
γ-Globulin, 69
Glues, 154
Glutamic acid, 84
Guinea pigs
   immunogenicity of chromium compounds in, 71—73
   maximization test in, 62

H

Hair analysis, 4
Hapten
   i.v. injection of, 98, 104

recognition of, 84
  with relatively long half-life, 105
Haptenized macrophages, 74, 84, 103
Hartley strains, 85
Hemolysis, 117
Heparin, 69
Heterologous antilymphocyte serum, 94
Hexavalent chromium, 2, 23, 46, 55, 56, 65—67, 73, 83, 84, 139, 142
  absorption of, 66
  binding capacity of, 68
  conversion of to trivalent forms, 69—71, 82
  immunogenicity of in guinea pig, 71—73
  penetration capacity of, 67
  protein denatured by, 69
  slowly-soluble, 36
  soluble, 32, 36
Hexavalent chromium compounds, 41, 43
Humans
  diets of, 14
  disease treatment with chromium in, 15
  epidemiologic studies, in, 15—23
Human serum albumin, 69
Humoral suppressor factors, 88
Hygiene, 19, 33
Hypersensitivity, 54, 57, 103
  chromium VI, 76

I

Ia-antigen, see Immune response gene associated antigen
Ia system, 71
Id system, 121
Immune complexes, 121
Immune deviation, 114
Immune response gene associated antigen (Ia-antigen), 84
Immunoblasts, 103
Immunogenicity
  chromium compounds in guinea pig, 71—73
  valence and, 64—84
Immunological mechanisms of contact sensitivity, 57—61
Immunological tolerance, 65, 88
  chromium compounds, 98—101
Immunosuppressive agents, 94—98
Implantation of intrabronchial pellets, 37, 46, 48, 49
Inactivation of macrophages, 88
Inbred strains, 84
Indirect MMT, 83, 101
Inductive phase, 58
Industrial dermatoses, 149
Infiltrate, 119
Inflammation of larynx, 4
Inflammation of liver, 4
Inflammatory lympholines, 122
Inhalation exposure, 24
Injection
  chromium, 4, 98

hapten, 98, 104
Insulin-requiring diabetics, 3
Internal absorption, 3
Intrabronchial pellet implantation, 37, 46, 48, 49
Intradermal testing, 99, 141
Intravenous injection of hapten, 98, 104
In-vitro reactions with contact sensitivity, 73—82
Iontophoretic application, 65
Irritation from chemicals, 32, 34

J

Job characterization, 18

K

Kidney, 4
Kinines, 121

L

Langerhans' cells, 59, 63, 67, 70, 84, 109, 117, 118
Laryngeal inflammation, 4
Latent period, 15—17, 19
Lead chromate, 21, 24, 25, 35, 56
Leather, 56, 152
Lesions, 61
Leukocytes
  antipolymorphonuclear, 114
  polymorphonuclear, 111, 114
Light sensitivity, 141
  associated, 146
Linear copolymers of tyrosine, 84
Linings of boilers, 155
Liver, 4
LNPF, see Lymph node permeability factor
Lobectomy, 37
Local resection, 37
Local Shwartzman phenomenon, 109
LTT, see Lymphocyte transformation test
Lung cancer, 14, 18, 32, 35—40, 45, 47
  early diagnosis of, 36—40
  epidemiology of, 35—40
Lymph node
  draining, 60
  lymphocytes of, 76
Lymph node permeability factor (LNPF), 114
Lymphocytes
  effector, 104
  lymph node, 76
  MIF-producing, 79
  peritoneal exudate, 74, 78, 104
  proliferative response of, 60
  specifically reactive, 88
  transfer of, 103
Lymphocyte transfer reaction, 118
Lymphocyte transformation test (LTT), 60, 73, 79—84

Lymphokines, 61, 78
  inflammatory, 122
Lysine, 84

M

Machine oils, 154—155
Macrophage migration inhibition test (MMT), 73, 77—79
  direct, 101
  indirect, 83, 101
Macrophages, 59, 67, 70
  chromium-modified, 81, 83
  haptenized, 74, 84, 103
  inactivation of, 88
  migration of, 78
Magnetic tapes, 155
Masking of tolerance, 106
Mass spectrometry, 6—8
Matches, 154
Maximization test for guinea pig, 62
Maximum allowable concentrations, 19
Mechanical lesions of skin, 61
Mediators, 119
Memory cells, 103, 104
Mercury chloride, 84, 86
Mercury compounds, 74
Metal contact sensitivity, 84—87
Metallic compounds, 74
Metal salts, 119
Methionine, 69
Methotrexate (MTX), 94—96
MIF, see Migration inhibition factor
Migration inhibition factor (MIF), see also Macrophage migration inhibition test, 73—74, 83, 101
  production of by lymphocytes, 79
Migration of macrophages, 78
Milk testers, 156
Minimum initiation dose, 15, 17
Mitomycin C, 74
Mixed chromates, 24
MMT, see Macrophage migration inhibition test
Molybdenum, 55
Monitoring of air, 33
Monochromates, 18—20, 25
Mononuclear cell infiltrate, 119
Mortality ratio, 23
MTX, see Methotrexate
Municipal drinking water supplies, 3, 14
Muscle, 4
  thigh, 46
Mutagenicity, 41—45
  of chromium, 18

N

Nail samples, 4
Nasal cancer, 19, 34
Nasal septum, 19
  perforation of, 4, 20, 32—34
  ulceration of, 32, 138
Natural waters, 3, 14
Neosalvarsan, 87, 89, 92, 103, 104, 108, 120, 121
  tolerance of during primary response, 99
Nickel, 52, 69
  reactions to, 55
Nickel sulfate, 74
Noncarcinogenic chromates, 18
Nonspecific flare-up, 120
Normal body exposure to chromium, 2
Normal lymphocyte transfer reaction, 118
Nose bleeding, 15

O

Observation time, 17
Observed/expected ratio, 19
Obstruction of airways, 32
Occupational cancer, 18
Occupational exposure, 4, 14
Occupational history, 18, 26
Oils, 154—155
One-bath method, 70
Optical spectroscopy, 5—6
Orally absorbed chromium, 4, 157—158
Ouchterlony, 117
Oxalate, 142
Oxazolon, 120

P

Paints, 149—150
Paper and wood industry, 154
Parabiosis, 103
Particle excitation, 8
Passivation, 150
Passive cutaneous anaphylaxis, 117
Patch testing, 141—144
Pellet implantation, 37, 46, 48, 49
Penetration
  chromium compounds, 65
  route of, 64
Penetration capacity, 64—67, 73, 82, 84
  hexavalent chromium compounds, 67
  skin, 69
  trivalent chromium compounds, 67
Perforation nasal septum, 4, 20, 32—34
Peritoneal exudate lymphocytes, 74, 78, 104
Permanent desensitization, 103, 104, 106, 112
Permeability factor, 114
Persistence of positive patch tests, 143
Personal hygiene, 19, 33
PHA, see Phytohemagglutinin
Phagocytosis, 70
Pharmacodynamic mediators, 119
pH dependence, 65
Phytohemagglutinin (PHA), 74
Picryl chloride, 88, 121
Pigments, 20—21, 25, 35, 152—153

Pirbright strains, 85
PIXE, see Proton induced X-ray emission
Plants, 14
    regulations of, 140
Poison ivy, 121
Polarography, 64
Poly-L-lysine, 84
Polymorphonuclear leukocytes, 111, 114
Polynuclear cell infiltrate, 119
Positive patch tests, 143
Potassium bichromate, 138
Potassium chromate, 56
Potassium dichromate, 56, 65, 68, 81, 84—86, 89—92, 99, 100, 103, 104, 109
    FCA and, 62, 75
    threshold concentration of, 55
Potassium oxalate, 56
Potentiometry, 9
Precipitation, 117
Precision, 8
Prednisolone, 98, 107
Preservatives for timber, 155
Pretreatment tolerance, 100, 105
Prevention
    chromium dermatitis, 157—158
    chromium ulcers, 139—140
Primary response tolerance, 99, 105
Primer paints, 149—150
Printing, 153
Production of chromium compounds, 18—21
Prognosis, 146
Proliferative response of lymphocytes, 60
Proportional mortality ratio, 23
Prostheses, 15
Protein carriers, 67, 69
Proteins
    binding capacity to, 67, 82
    covalent bonds with, 69
    denatured, 69
Proton induced X-ray emission (PIXE), 7
Pulmonary adenomata, 45

R

Radioactive sodium chromate, 66
Radiochromates, 15
Radioimmunoelectrophoresis, 117
Rash, 120—122
Reactions
    antibody-mediated, 58
    Arthus, 111, 114, 116, 118, 120
    cell-mediated, 58
    cobalt, 55
    flare-up, see Flare-up reactions
    id, 121
    in-vitro, 73—82
    lymphocyte transfer, 118
    nickel, 55
    primary, 99, 105
    re-test, 108
    secondary flare-up, 109—120
    skin, see Skin reactions
    spontaneous, 109, 120
    subliminal, 110, 111
    types of, 142
Receptors
    blockade of, 100
    specific, 104
Reducing agents, 140
Regulations of chrome plating, 140
Repeated epicutaneous application, 89
Repellant, 152
Replacement prostheses, 15
Research needs, 26
Resections, 37
Response, see Reactions
Re-test reaction, 108
Rhinorrhea, 32

S

Safe levels, 26
Salmonella/microsome test, 41
Salts, 87
    hexavalent, 55—67
    metal, 119
    solubility of, 56
    trivalent, 3, 55—67
Sand, 154
Sarcoma, see also Cancer, 24
    of lung, 45
Sea water, 2, 3, 14
Secondary flare-up, 109—120
Secondary phenomena during contact sensitivity, 107—122
Second-shot challenge, 105, 106
Sensitivity
    contact, see Contact sensitivity
    light, see Light sensitivity
Sensitization, 55—57, 140
Serum
    antilymphocyte, see Antilymphocyte serum
    antipolymorphonuclear, 114, 115
    antithymocyte, 94
Serum albumin
    bovine, 74
    human, 69
Serum γ-globulin, 69
Serum values, 4
Sex incidence of chromate sensitivity, 144
SH, see Sulfhydril
Shoe dermatitis, 152, 153
Shwartzman phenomenon, 120
    local, 109
Site of testing, 142
Skin, 4
    chrome ulceration of, 34
    mechanical lesions of, 61
    penetration capacity into, 69
Skin extracts, 69
Skin reactions
    antibody-mediated, 58

cell-mediated, 58
chromate sensitization, 140
drug, 109
Skin reactive factor (SRF), 73—77
Slowly-soluble hexavalent chromium, 36
Smoking habits, 18, 20, 39
SMR, see Standardized mortality ratio
Sodium calcium edetate, 34
Sodium chromate, 56, 65
   radioactive, 66
Sodium dichromate, 56, 150, 152
Sodium dithionite, 140
Sodium lauryl sulfate, 62
Sodium metabisulfite, 140
Soil, 3
Solubility
   chromium salt, 56
   water, 25
Soluble hexavalent chromium compounds, 32
Solvents, 151
Sources of exposure, 33
Spatial distribution of elementation composition, 9
Specifically reactive lymphocytes, 88
Specificity of carrier, 59
Specific receptors, 104
Spectroscopy, 9, 64
   atomic absorption, 5, 9
   flame emission, 9
   mass, 6—8
   optical, 5—6
Split adjuvant technique, 62
Split tolerance, 114, 116
Spontaneous reactions, 109, 120
Sputum cytology, 37
Squamous carcinoma, 45
   in lungs, 47
SRF, see Skin reactive factor
Stainless steel welding, 22—23
Standardized mortality ratio (SMR), 23
Stimulator cells, 74, 87, 118
Strains
   Hartley, 85
   inbred, 84
   Pirbright, 85
Subliminal reaction, 110, 111
Sufficient dose, 15
Sugar, 3
Sulfate, 56
Sulfhydril (SH) groups, 69
Suppressor cells, 61, 88, 105, 106
   activation of, 88
Suppressor factors, 88
Synthetic antigens, 84, 86, 87

T

Tanning, 152
   capacity of, 70
   leather, 152
   liquor for, 53

T cells, 58, 84, 87
Television workers, 155
Testing, see also specific tests, 55—57
   site of, 142
Thigh muscle fibrosarcoma, 46
Threshold concentrations, 55
Threshold limit values, 19
Timber preservatives, 155
Tire fitters, 155
Tissue levels, 3—4
Tolerance, 65, 103
   chromium contact sensitivity, 87—107
   DNCB, 99
   DNCB contact sensitivity, 103
   immunological, 65, 88, 98—101
   masking of, 106
   mechanisms of, 101—107
   pretreatment, 100, 105
   primary response and, 99, 105
   split, 114, 116
   transfer of, 103
Trace elements
   analysis of, 5
   biological dose response to, 2
Transfer reaction, 118
Treatment
   chrome ulcers, 140
   chromium in, 15
Triton X—100, 62
Trivalent chromium, 2, 3, 20, 23, 32, 41, 43, 44, 55, 56, 66, 70, 73, 83, 139, 142
   binding capacity of, 68
   conjugates of, 69
   conversion of hexavalent form to, 69—71, 82
   immunogenicity of in guinea pig, 71—73
   penetration capacity of, 67
Trivalent chromium sulfate, 92
Trypsin, 104
Tumors
   bronchial, 24
   induction of in experimental animals, 45—49
Tungsten, 55
Two-bath procedure, 70
Tyrosine, 84

U

Ulceration, 32—34, 138—140
   nasal, 32, 138
   penetrating, 4
   skin, 34
Users of chromium compounds, 21—23

V

Valence, 82—84, 140
   immunogenicity and, 64—84
   significance of, 55—57
Vasoactive kinines, 121
Volumetry, 9

W

Water
    municipal drinking, 3, 14
    natural, 3, 14
    sea, 2, 3, 14
Water repellant, 152
Water solubility, 25
Welding, 151—152
    fumes from, 22, 25
    stainless steel, 22—23
Wood and paper industry, 154

X

X-ray emission, 7—8
    proton-induced, 7
X-ray fluorescence, 8

Z

Zinc chromate, 20, 21, 24, 25, 35, 150
Zinc potassium chromate, 48, 149
Zinc tetraoxy-chromate, 150